SCHAUM'S™
EASY OUTLINES

Beginning Chemistry

Online Diagnostic Test

Go to **Schaums.com** to launch the Schaum's Diagnostic Test.

This convenient application provides a 30-question multiple-choice test that will pinpoint areas of strength and weakness to help you focus your study. Questions cover all aspects of beginning chemistry, and the correct answers are explained in full. With a question-bank that rotates daily, the Schaum's Online Test also allows you to check your progress and readiness for final exams.

Other titles featured in Schaum's Online Diagnostic Test:

Schaum's Easy Outlines: Geometry, 2nd Edition
Schaum's Easy Outlines: Calculus, 2nd Edition
Schaum's Easy Outlines: Statistics, 2nd Edition
Schaum's Easy Outlines: Elementary Algebra, 2nd Edition
Schaum's Easy Outlines: College Algebra, 2nd Edition
Schaum's Easy Outlines: Biology, 2nd Edition
Schaum's Easy Outlines: Human Anatomy and Physiology, 2nd Edition
Schaum's Easy Outlines: Organic Chemistry, 2nd Edition
Schaum's Easy Outlines: College Chemistry, 2nd Edition

Beginning Chemistry

Second Edition

David Goldberg, Ph.D.

Abridgement Editor:
Katherine E. Cullen, Ph.D.

New York Chicago San Francisco Lisbon London Madrid Mexico City
Milan New Delhi San Juan Seoul Singapore Sydney Toronto

The McGraw·Hill Companies

4 5 6 7 8 9 10 11 12 13 14 15 QFR/QFR 1 9 8 7 6 5 4 3

ISBN 978-0-07-174588-8
MHID 0-07-174588-2

Library of Congress Cataloging-in-Publication Data

Goldberg, David E. (David Elliott), 1932-
Schaum's easy outline of beginning chemistry / David E. Goldberg. — 2nd ed.
p. cm. — (Schaum's outlines)
Includes index.
ISBN 0-07-174588-2 (alk. paper)
1. Chemistry—Outlines, syllabi, etc. 2. Chemistry—Problems, exercises, etc. I. Title II. Title: Easy outline of beginning chemistry.

QD41.G648 2010
540—dc22 2010010888

This book is printed on acid-free paper.

Contents

Chapter 1
BASIC CONCEPTS

IN THIS CHAPTER:

- ✔ *The Elements*
- ✔ *Matter and Energy*
- ✔ *Properties*
- ✔ *Classification of Matter*
- ✔ *Representation of Elements*
- ✔ *Laws, Hypotheses, and Theories*
- ✔ *Solved Problems*

The Elements

Chemistry is the study of matter and energy and the interaction between them. The **elements** are the building blocks of all types of matter in the universe. An element cannot be broken down into simpler substances by ordinary means. A few more than 100 elements and the many combinations of these elements account for all the materials of the world.

The elements occur in widely varying quantities on earth. The 10 most abundant elements make up 98 percent of the mass of the crust of the earth. Many elements occur only in traces, and a few elements are synthetic. The elements are not distributed uniformly throughout the earth. The distinct properties of the different elements cause them to be concentrated more or less, making them available as raw materials.

Matter and Energy

Chemistry is the study of matter, including its composition, its properties, its structure, the changes which it undergoes, and the laws governing those changes. **Matter** is anything that has mass and occupies space. Any material object, no matter how large or small, is composed of matter. In contrast, light, heat, and sound are forms of energy. **Energy** is the ability to produce change. Whenever a change of any kind occurs, energy is involved; and whenever any form of energy is changed to another form, it is evidence that a change of some kind is occurring or has occurred.

The concept of mass is central to the discussion of energy. The **mass** of an object depends on the quantity of matter in the object. The more mass an object has, the more it weighs, the harder it is to set in motion, and the harder it is to change the object's velocity once it is in motion.

Matter and energy are now known to be interconvertible. The quantity of energy producible from a quantity of matter, or vice versa, is given by Einstein's famous equation

$$E = mc^2$$

where E is the energy, m is the mass of the matter which is converted to energy, and c^2 is a constant—the square of the velocity of light. The constant c^2 is so large,

$$90{,}000{,}000{,}000{,}000{,}000 \text{ m}^2/\text{s}^2 \text{ or } 34{,}600{,}000{,}000 \text{ mi}^2/\text{s}^2$$

that tremendous quantities of energy are associated with conversions of minute quantities of matter to energy. The quantity of mass accounted for by the energy contained in a material object is so small that it is not measurable. Hence, the mass of an object is very nearly identical to the quantity of matter in the object. Particles of energy have very small masses despite having no matter whatsoever; that is, all the mass of a particle of light is associated with its energy.

The mass of an object is directly associated with its weight. The **weight** of a body is the pull on the body by the nearest celestial body. On earth, the weight of a body is the pull of the earth on the body, but on the moon, the weight corresponds to the pull of the moon on the body. The weight of a body is directly proportional to its mass and also depends on the distance of the body from the center of the earth or moon or whatever celestial body the object is near. In contrast, the mass of an object is independent of its position. For example, at any given location on the surface of the earth, the weight of an object is directly proportional to its mass.

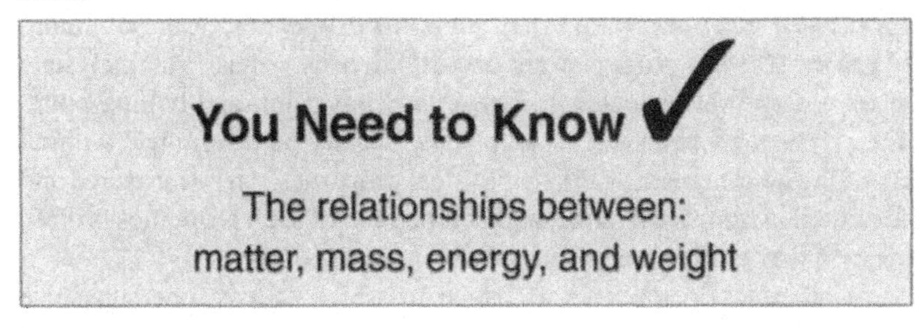

Since the study of chemistry is concerned with the changes that matter undergoes, chemistry is also concerned with energy. Energy occurs in many forms—heat, light, sound, chemical energy, mechanical energy, electrical energy, and nuclear energy. In general, it is possible to convert each of these forms of energy to others. Except for reactions in which the quantity of matter is changed, as in nuclear reactions, the **law of conservation of energy** is obeyed. In fact, many chemical reactions are carried out for the sole purpose of producing energy in a desired form. For example, in the burning of fuels in homes, chemical energy is converted to heat; in the burning of fuels in automobiles, chemical energy is converted to mechanical energy of motion.

The Law of Conservation of Energy:

Energy can neither be created nor destroyed (in the absence of nuclear reactions).

Properties

Every substance has certain characteristics that distinguish it from other substances and that may be used to establish that two specimens of the same substance are indeed the same. Those characteristics that serve to distinguish and identify a specimen of matter are called the **properties** of the substance. The properties related to the **state** (gas, liquid, or solid) or appearance of a sample are called **physical properties**. Some commonly known physical properties are **density** (density = mass/volume), state at room temperature, color, hardness, melting point, and boiling point. The physical properties of a sample can usually be determined without changing its composition. Many physical properties can be measured and described in numerical terms, and comparison of such properties is often the best way to distinguish one substance from another.

A chemical reaction is a change in which at least one substance changes its composition and its set of properties. The characteristic ways in which a substance undergoes chemical reaction, or fails to undergo chemical reaction are called its **chemical properties**. Examples of chemical properties are flammability, rust resistance, reactivity, and biodegradability.

Classification of Matter

To study the vast variety of materials that exist in the universe, the study must be made in a systematic manner. Therefore, matter is classified according to several different schemes. Matter may be classified as organic or inorganic. It is **organic** if it is a compound of carbon and hydrogen (see Chapter 14). Otherwise, it is **inorganic**. Another such scheme uses the composition of matter as a basis for classification; other schemes are based on chemical properties of the various classes. For examples, substances may be classified as acids, bases, or salts. Each scheme is useful, allowing the study of a vast variety of materials in terms of a given class.

In the method of classification of matter based on composition, a given specimen of material is regarded as either a pure substance or a mixture. The term **pure substance** refers to a material all parts of which have the same composition and that has a definite and unique set of properties.

In contrast, a **mixture** consists of two or more substances and has a somewhat arbitrary composition. The properties of a mixture are not unique, but depend on its composition. The properties of a mixture tend to reflect the properties of the substances of which it is composed; that is, if the composition is changed a little, the properties will change a little.

There are two kinds of substances—elements and compounds. **Elements** are substances that cannot be broken down into simpler substances by ordinary chemical means. Elements cannot be made by the combination of simpler substances. There are slightly more than 100 elements, and every material object in the universe consists of one or more of these elements.

Remember

Familiar substances that are elements include carbon, aluminum, iron, copper, gold, oxygen, and hydrogen.

Compounds are substances consisting of two or more elements combined in definite proportions by mass to give a material having a definite set of properties different from that of any of its constituent elements. For example, the compound water consists of 88.8 percent oxygen and 11.2 percent hydrogen by mass. The physical and chemical properties of water are distinctly different from those of both hydrogen and oxygen. For example, water is a liquid at room temperature and pressure, while the elements of which it is composed are gases under these same conditions. Chemically, water does not burn; hydrogen may burn explosively in oxygen (or air). Any sample of pure water, regardless of its source, has the same composition and the same properties.

There are millions of known compounds, and thousands of new ones are discovered or synthesized each year. Despite such a vast number of compounds, it is possible for the chemist to know certain properties of each one, because compounds can be classified according to their composition and structure, and groups of compounds in each class have some properties in common. For example, organic compounds are generally combustible in oxygen, yielding carbon dioxide and water. So, any compound that contains carbon and hydrogen may be predicted by the chemist to be combustible in oxygen.

There are two kinds of mixtures—homogeneous mixtures and heterogeneous mixtures. **Homogeneous mixtures** are also called **solutions**, and **heterogeneous mixtures** are sometimes called simply mixtures. In heterogeneous mixtures, it is possible to see differences in the sample, merely by looking, although a microscope may be required. In contrast, homogeneous mixtures look the same throughout the sample, even under the best optical microscope.

Representation of Elements

Each element has an internationally accepted symbol to represent it. The periodic table at the back of this book includes both the names and symbols of the elements. Note that symbols for most elements are merely abbreviations of their names, consisting of either one or two letters. Three letter symbols are used for elements over number 103. The first letter of the symbol is always written as a capital letter; the second and third letters, if any, are written as lowercase letters. The symbols of a few elements do not suggest their English names, but are derived from the Latin or German names of the elements.

You Need to Memorize ✔

The names and symbols of the common elements.

A convenient way of displaying the elements is in the form of a **periodic table**, such as is shown at the end of this book. The basis for the arrangement of the elements in the periodic table will be discussed more in Chapter 2.

Laws, Hypotheses, and Theories

A statement that generalizes a quantity of experimentally observable phenomena is called a **scientific law**. For example, if a person drops a pencil, it falls downward. This result is predicted by the law of gravity. A generalization that attempts to explain why certain experimental results occur is called a **hypothesis**. When a hypothesis is accepted as true by the scientific community, it is then called a **theory**. One of the most important scientific laws is the **law of conservation of mass**: During any process (chemical reaction, physical change, or even nuclear reaction) mass is neither created nor destroyed. Because of the close approximation that the mass of an object is the quantity of matter it contains (excluding the mass corresponding to its energy), the law of conservation of mass can be approximated by the **law of conservation of matter**: During an ordinary chemical reaction, matter can be neither created nor destroyed.

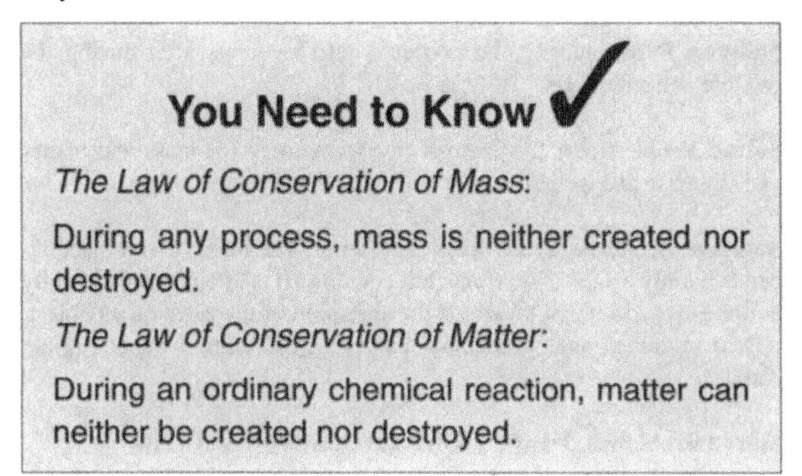

You Need to Know

The Law of Conservation of Mass:

During any process, mass is neither created nor destroyed.

The Law of Conservation of Matter:

During an ordinary chemical reaction, matter can neither be created nor destroyed.

Solved Problems

Solved Problem 1.1 TNT is a compound of carbon, nitrogen, hydrogen, and oxygen. Carbon occurs in many forms—including graphite (the material in "lead pencils") and diamond. Oxygen and nitrogen comprise over 98 percent of the atmosphere. Hydrogen is an element that reacts explosively with oxygen. Which of the properties of the elements determines the properties of TNT?

Solution: The properties of the elements do not matter. The properties of the compound are quite independent of those of the elements. A compound has its own distinctive set of properties. TNT is most noted for its explosiveness.

Solved Problem 1.2 Name an object or instrument that changes (*a*) electrical energy to light, (*b*) motion to electrical energy, (*c*) chemical energy to heat, and (*d*) chemical energy to electrical energy.

Solution: (*a*) light bulb, (*b*) generator or alternator, (*c*) gas stove, and (*d*) battery.

Solved Problem 1.3 A teaspoon of salt is added to a cup of warm water. White crystals are seen at the bottom of the cup. Is the mixture homogeneous or heterogeneous? Then the mixture is stirred until the salt crystals disappear. Is the mixture now homogeneous or heterogeneous?

Solution: Before stirring, the mixture is heterogeneous; after stirring, the mixture is a solution.

Solved Problem 1.4 Distinguish clearly between (*a*) mass and matter and (*b*) mass and weight.

Solution: (*a*) Matter is any kind of material. The mass of an object depends mainly on the matter which it contains. It is affected only slightly by the energy in it. (*b*) Weight is the attraction of the earth on an object. It depends on the mass of the object and the distance to the center of the earth.

Solved Problem 1.5 Distinguish between a theory and a law.

Solution: A law tells what happens under a given set of circumstances, while a theory attempts to explain why that behavior occurs.

IN THIS CHAPTER:

- ✔ *Atomic Theory*
- ✔ *Atomic Masses*
- ✔ *Atomic Structure*
- ✔ *Isotopes*
- ✔ *Periodic Table*
- ✔ *Solved Problems*

Atomic Theory

In 1804, John Dalton proposed the existence of atoms. He not only postulated that atoms exist, as had ancient Greek philosophers, but he also attributed certain properties to the atom. His postulates were as follows:

1. Elements are composed of indivisible particles, called **atoms**.
2. All atoms of a given element have the same mass, and the mass of an atom of a given element is different from the mass of an atom of any other element.
3. When elements combine to form compounds, the atoms of one element combine with those of the other element(s) to form **molecules**.
4. Atoms of two or more elements may combine in different ratios to form different compounds.

5. The most common ratio of atoms is 1:1, and where more than one compound of two or more elements exists, the most stable is the one with 1:1 ratio of atoms. (This postulate is incorrect.)

Dalton's postulates stimulated great activity among chemists, who sought to prove or disprove them. The fifth postulate was very quickly shown to be incorrect, and the first three have had to be modified in light of later knowledge. However, the first four postulates were close enough to the truth to lay the foundations for a basic understanding of mass relationships in chemical compounds and chemical reactions.

Dalton's postulates were based on three laws that had been developed shortly before he proposed his theory.

1. The **law of conservation of mass** (see Chapter 1) states that mass is neither created nor destroyed in a chemical reaction.

2. The **law of definite proportions** states that every chemical compound is made up of elements in a definite ratio by mass.

3. The **law of multiple proportions** states that when two or more different compounds are formed from the same elements, the ratio of masses of each element in the compounds for a given mass of any other element is a small whole number.

Dalton argued that these laws are entirely reasonable if the elements are composed of atoms. For example, the reason that mass is neither gained nor lost in a chemical reaction is that the atoms merely change partners with one another; they do not appear or disappear. The definite proportions of compounds stem from the fact that the compounds consist of a definite ratio of atoms, each with a definite mass. The law of multiple proportions is due to the fact that different numbers of atoms of one element can react with a given number of atoms of a second element, and since the atoms must combine in whole-number ratios, the ratio of the masses must also be in whole numbers.

Atomic Masses

Once Dalton's hypotheses had been proposed, the next logical step was to determine the relative masses of the atoms of the elements. Since there was no way at the time to determine the mass of an individual atom, the relative masses were the best information available. That is, one could tell that an atom of one element had a mass twice as great as an atom of a different element (or 15/4 as much, or 17.3 times as much, etc.). The relative masses could be determined by taking equal (large) numbers of atoms of two elements and determining the ratio of masses of these collections of atoms.

For example, a large number of carbon atoms have a total mass of 12.0 g, and an equal number of oxygen atoms have a total mass of 16.0 g. Since the number of atoms of each kind is equal, the ratio of the masses of one carbon atom to one oxygen atom is 12.0 to 16.0. One ensures equal numbers of carbon and oxygen atoms by using a compound of carbon and oxygen in which there are equal numbers of the two elements (i.e., carbon monoxide).

A great deal of difficulty was encountered at first, because Dalton's fifth postulate gave an incorrect ratio of numbers of atoms in many cases. Such a large number of incorrect results were obtained that it soon became apparent that the fifth postulate was not correct. It was not until some 50 years later that an experimental method was devised to determine the atomic ratios in compounds, at which time the scale of relative atomic masses was determined in almost the present form. These relative masses are called **atomic masses**, or sometimes **atomic weights**.

The atomic mass of the lightest element, hydrogen, was originally taken to be one **atomic mass unit** (amu). The modern values of the atomic masses are based on the most common kind of carbon atom, called "carbon-12" and written ^{12}C, as the standard. The mass of ^{12}C is measured in the modern **mass spectrometer**, and ^{12}C is defined to have an atomic mass of exactly 12 amu. On this scale, hydrogen has an atomic mass of 1.008 amu.

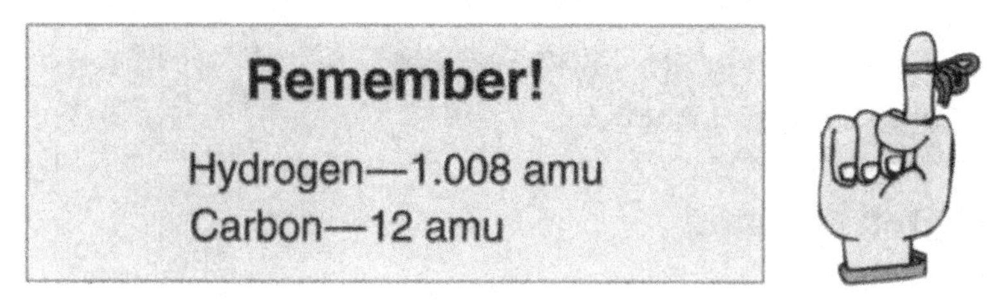

Different names are used for the unit of atomic mass by different authors, and different abbreviations are used for it. The term **Dalton** is used by some, abbreviated D. Other authors use amu. The abbreviation u rather than amu is sometimes encountered.

The atomic mass of an element is the relative mass of an average atom of the element compared with ^{12}C, which has a mass of exactly 12 amu. Thus, since a sulfur atom has a mass 8/3 times that of a carbon atom, the atomic mass of sulfur is

$$12 \text{ amu} \times 8/3 = 32 \text{ amu}$$

The modern values of the atomic masses of the elements are given in the periodic table.

Atomic Structure

From 50 years to 100 years after Dalton proposed his theory, various discoveries showed that the atom is not indivisible, but is really composed of parts. Natural radioactivity and the interaction of electricity with matter are two different types of evidence for this subatomic structure. The most important subatomic particles are listed in Table 2.1, along with their most important properties. The **protons** and **neutrons** occur in a very tiny **nucleus**. The **electrons** occur outside the nucleus.

	Charge (*e*)	Mass (amu)	Location
Proton	+1	1.00728	In nucleus
Neutron	0	1.00894	In nucleus
Electron	−1	0.0005414	Outside nucleus

Table 2.1 Subatomic particles

There are two types of electric charges that occur in nature, positive and negative. Charges of these two types are opposite one another and cancel the effect on one another. Bodies with opposite charge types attract one another; those with the same charge type repel one another. If a body has equal numbers of charges of the two types, it has no net charge and is said to be **neutral**. The charge on the electron is a fundamental unit of electric charge (equal to 1.6×10^{-19} C) and is given the symbol e.

The number of protons in the nucleus determines the chemical properties of the element. That number is called the **atomic number** of the element. Each element has a different atomic number. An element may be identified by giving its name or its atomic number. Atomic numbers may be specified by use of a subscript before the symbol of the element. For example, carbon may be designated $_6C$. Atomic numbers are included in the periodic table.

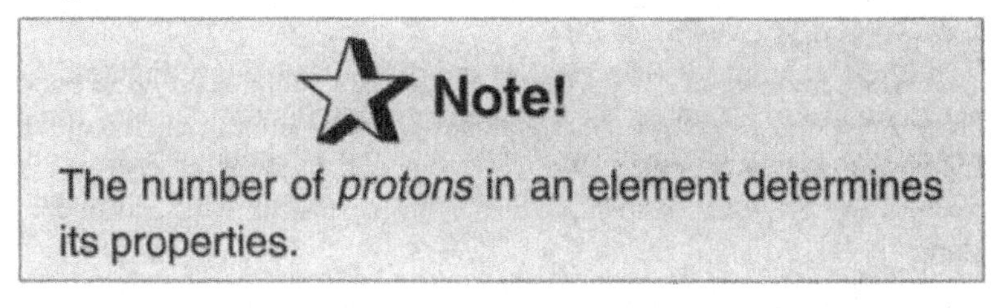

Isotopes

Atoms having the same number of protons but different numbers of neutrons are called **isotopes** of one another. The number of neutrons does not affect the chemical properties of atoms appreciably, so all isotopes of a given element have essentially the same chemical properties. Different isotopes have different masses and different nuclear properties, however.

The sum of the number of protons and the number of neutrons in the isotope is called the **mass number** of the isotope. Isotopes are usually distinguished from one another by their mass numbers, given as a superscript before the chemical symbol for the element. Carbon-12 is an isotope of carbon with a symbol ^{12}C.

Periodic Table

The **periodic table** (see Appendix) is a very useful tabulation of the elements. It is constructed so that each vertical column contains elements

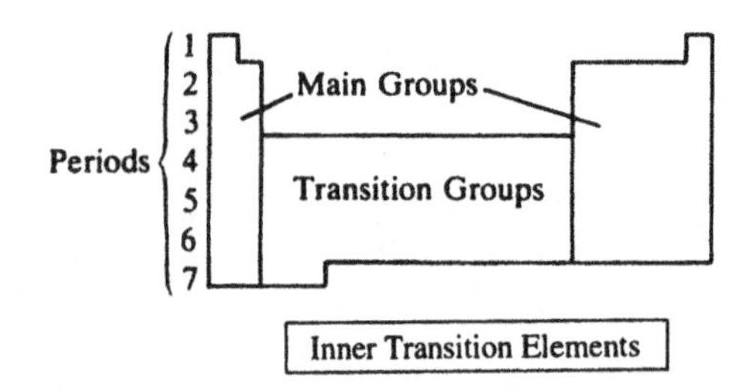

Figure 2-1 Periods and element types in the periodic table

that are chemically similar. The elements in the columns are called **groups**, or **families**. Elements in some groups are very similar to one another. Elements in other groups are less similar. Each row in the table is called a **period** (see Figure 2-1).

There are three distinct areas of the periodic table (see Figure 2-1): the **main group elements**, the **transition group elements**, and the **inner transition group elements**. We will focus our attention at first on the main group elements, whose properties are easiest to learn and understand.

The periods and the groups are identified differently. The periods are labeled from 1 to 7. Some reference is made to the period numbers. The groups are referred to extensively by number. Unfortunately, the groups have been labeled in three different ways:

1. **Classical**—Main groups are labeled IA through VIIA plus 0. Transition groups are labeled IB through VIII (although not in that order).
2. **Amended**—Main groups and transition groups are labeled IA through VIII and then IB through VIIB plus 0.
3. **Modern—**Groups are labeled with Arabic numerals from 1 through 18.

Although the modern designation is seemingly simpler, it does not emphasize some of the relationships that the older designations do. In this book, the classical system will be followed, with the modern number often included in parentheses.

Several important groups are given names. Group IA (1) metals (not including hydrogen) are called the **alkali metals**. Group IIA (2) elements are known as the **alkaline earth metals**. Group VIIA (17) elements are called the **halogens**. Group IB (11) metals are known as the **coinage metals**. Group 0 (18) elements are known as the **noble gases**. These names lessen the need for using group numbers and thereby lessen the confusion from different systems.

The electrons are arranged in shells. (A more detailed account of electronic structure will be given in Chapter 3.) The maximum number of electrons that can fit in any shell n is given by

$$\text{Maximum number} = 2n^2$$

Since there are only about 100 electrons total in even the biggest atoms, the shells numbered 5 or higher never get filled with electrons. Another important limitation is that the outermost shell, called the **valence shell**, can never have more than eight electrons in it. The number of electrons in the valence shell is a periodic property.

Shell number	1	2	3	4	5	6	7
Maximum number of electrons	2	8	18	32	50	72	98
Maximum number as outermost shell	2	8	8	8	8	8	8

The number of outermost electrons is crucial to the chemical bonding of the atom (see Chapter 4). For the main group elements, the number of outermost electrons is equal to the classical group number, except that it is 2 for helium and 8 for the other group 0 (18) elements.

Solved Problems

Solved Problem 2.1 Find the charge on a nucleus which contains (*a*) 19 protons and 20 neutrons and (*b*) 19 protons and 22 neutrons.

Solution: (*a*) $19(^+1) + 20(0) = {}^+19$ and (*b*) $19(^+1) + 22(0) = {}^+19$

Both nuclei have the same charge. Although the nuclei have different numbers of neutrons, the neutrons have no charges, so they do not affect the charge on the nucleus.

Solved Problem 2.2 What is the charge on a boron nucleus? What is the charge on a boron atom?

Solution: The charge on a boron nucleus is $^{+}5$, but the charge on a boron atom is 0. Note that these questions sound very much alike, but are very different.

Solved Problem 2.3 (*a*) What is the sum of the number of protons and the number of neutrons in ^{12}C? (*b*) What is the number of protons in ^{12}C? (*c*) What is the number of neutrons in ^{12}C?

Solution: (*a*) 12, its mass number. (*b*) 6, its atomic number, given in the periodic table. (*c*) $12 - 6 = 6$.

Solved Problem 2.4 What is the maximum number of electrons in the third shell of an atom in which there are electrons in the fourth shell?

Solution: The maximum number of electrons in the third shell is 18.

Solved Problem 2.5 Arrange the 11 electrons of sodium (Na) into shells.

Solution: The first two electrons fill the first shell, and the next eight fill the second shell. That leaves one electron in the third shell.

Solved Problem 2.6 If 50 lb of coal burns into 5 oz of ashes, how is the law of conservation of mass obeyed?

Solution: The coal plus oxygen has a certain mass. The ashes plus the carbon dioxide (and perhaps a few other compounds) must have a combined mass that totals the same as the combined mass of the coal and oxygen. The law does not state that the total mass before and after the reaction must be the mass of the solids only.

Solved Problem 2.7 If 50.54 percent of naturally occurring bromine atoms have a mass of 78.9183 amu and 49.46 percent have a mass of 80.9163 amu, calculate the atomic mass of bromine.

Solution: $\left(\frac{50.54}{100}\right)(78.9183 \text{ amu}) + \left(\frac{49.46}{100}\right)(80.9163 \text{ amu}) = 79.91 \text{ amu}$

Chapter 3

ELECTRONIC CONFIGURATION OF THE ATOM

IN THIS CHAPTER:

- ✔ *Bohr Theory*
- ✔ *Quantum Numbers and Electron Energies*
- ✔ *Shells, Subshells, and Orbitals*
- ✔ *Orbital Shape*
- ✔ *Buildup Principle*
- ✔ *Electronic Structure and the Periodic Table*
- ✔ *Solved Problems*

Bohr Theory

The first plausible theory of the electronic structure of the atom was proposed in 1914 by Niels Bohr, a Danish physicist. To explain the hydrogen spectrum, he suggested that in each hydrogen atom the electron revolves about the nucleus in one of several possible circular orbits, each having a definite radius corresponding to a definite energy for the electron. An electron in the orbit closest to the nucleus should have the lowest energy. With the electron in that orbit, the atom is said to be in its lowest energy state, or **ground state**. If a discrete quantity of additional energy were absorbed by the atom, the electron might be able to move into another orbit having a higher energy. The hydrogen atom would then be in an **excited state**. An atom in the excited state will return to the ground and give off its excess energy as light in the process. In returning to the ground state, the energy may be emitted all at once, or it may be emitted in a stepwise manner. Since each orbit corresponds to a definite energy level, the energy of the light emitted will correspond to the definite differences in energy between levels. Therefore, the light emitted as the atom returns to its ground state will have a definite energy or a definite set of energies. The discrete amounts of energy emitted or absorbed by an atom or molecule are called **quanta** (singular, **quantum**). A quantum of light energy is called a **photon**.

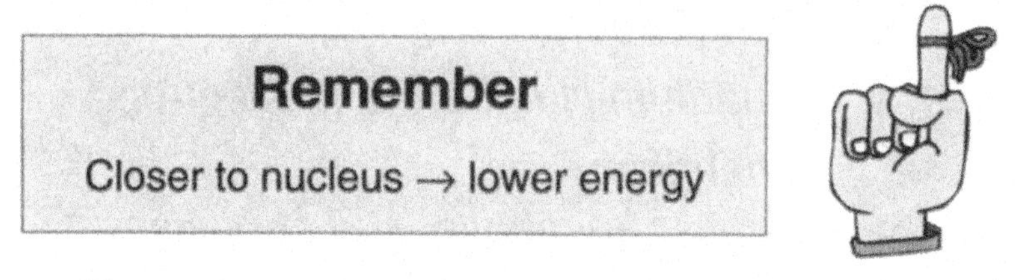

The wavelength of a photon is inversely proportional to the energy of the light, and when the light is observed through a spectroscope, lines of different colors, corresponding to different wavelengths, are seen.

Bohr's original idea of orbits of discrete radii has been greatly modified, but the concept that the electron in the hydrogen atom occupies different energy levels still applies. The successive energy levels are referred to as electron **shells**. The shells are sometimes designated by capital letters, with K denoting the lowest energy level, as follows:

Energy level:	1	2	3	4	5	...
Shell notation:	K	L	M	N	O	...

The electrons in atoms other than hydrogen also occupy various energy levels. The maximum number of electrons that can occupy a given shell depends on the shell number. For example, in any atom, the first shell can hold a maximum of only two electrons, the second shell can hold a maximum of eight electrons, the third shell can hold a maximum of 18 electrons, and so forth. The maximum number of electrons that can occupy any particular shell is $2n^2$, where n is the shell number.

Quantum Numbers and Electron Energies

A given electron is specified in terms of four quantum numbers that govern its energy, its orientation in space, and its possible interactions with other electrons. Thus, listing the values of the four quantum numbers describes the probable location of the electron, somewhat analogously to listing the section, row, seat and date on a ticket to a football game. To learn to express the electronic structure of an atom, it is necessary to learn: (1) the names, symbols, and permitted values of the quantum numbers (see Table 3.1) and (2) the order of increasing energy of electrons as a function of their sets of quantum numbers.

Name of Quantum Number	Symbol	Limitations
Principal	n	Any positive integer
Angular momentum	l	$0, \ldots, n-1$ in integer steps
Magnetic	m	$-l, \ldots, 0, \ldots, +l$ in integer steps
Spin	s	$-\frac{1}{2}, +\frac{1}{2}$

Table 3.1 Quantum numbers

The **principal quantum number** of an electron is the most important quantum number in determining the energy of the electron. In general, the higher the principal quantum number, the higher the energy of the electron. Electrons with higher principal quantum numbers are also apt to be farther away from the nucleus than electrons with lower principal quantum numbers. The first seven principal quantum numbers are the only important ones for electrons in ground states of atoms.

The **angular momentum quantum number** also affects the energy of an electron, but in general not as much as the principal quantum number does. In the absence of an electric or magnetic field around the atom, only these two quantum numbers have any effect on the energy of the electron. The values of l are often given letter designations, so that when they are stated along with principal quantum numbers, less confusion results. The letter designations of importance in the ground states of atoms are presented in Table 3.2.

l Value	Letter Designation
0	s
1	p
2	d
3	f

Table 3.2 Letter designations of angular momentum quantum numbers

The **magnetic quantum number** determines the orientation in space of the electron, but does not ordinarily affect the energy of an electron. Its values depend on the value of l for that electron.

The **spin quantum number** is related to the "spin" of the electron on its "axis." It ordinarily does not affect the energy of the electron.

The n and l quantum numbers determine the energy of each electron (apart from the effects of external electric and magnetic fields, which are most often not of interest in general chemistry courses). The energies of the electrons increase as the sum $n + l$ increases; the lower the value of $n + l$ for an electron in an atom, the lower its energy. For two electrons with equal values of $n + l$, the one with the lower n value has lower energy. Thus, we can fill an atom with electrons starting with its lowest-energy electrons by starting with the electrons with the lowest sum $n + l$.

The **Pauli exclusion principle** states that no two electrons in the same atom can have the same set of four quantum numbers. Along with the order of increasing energy, we can use this principle to deduce the order of filling of electron shells in atoms.

Shells, Subshells, and Orbitals

Electrons having the same value of n in an atom are said to be in the same **shell**. Electrons having the same value of n and the same value of l in an atom are said to be in the same **subshell**. The number of subshells within a given shell is merely the value of *n*, the shell number. Thus, the first shell has one subshell, the second shell has two subshells, and so forth. These facts are summarized in Table 3.3. Even the atoms with the most electrons do not have enough electrons to completely fill the highest shells shown. The subshells that hold electrons in the ground states of the biggest atoms are in boldface type.

Energy Level *n*	Type of Subshell	Number of Subshells
1	***s***	1
2	***s, p***	2
3	***s, p, d***	3
4	***s, p, d, f***	4
5	***s, p, d, f***, *g*	5
6	***s, p, d***, *f, g, h*	6
7	***s, p***, *d, f, g, h, i*	7

Table 3.3 Arrangement of subshells in electron shells

Depending on the permitted values of the magnetic quantum number *m*, each subshell is further broken down into units called orbitals. The number of orbitals per subshell depends on the type of subshell, but not on the value of *n*. Each orbital can hold a maximum of two electrons; hence, the maximum number of electrons that can occupy a given subshell is determined by the number of orbitals available. These relationships are presented in Table 3.4. The maximum number of electrons in any given energy level is thus determined by the subshells it contains. The first shell can contain two electrons; the second, 8 electrons; the third, 18 electrons; the fourth 32 electrons; and so on.

Type of Subshell	Allowed Values of *m*	Number of Orbitals	Maximum number of electrons
s	0	1	2
p	-1, 0, 1	3	6
d	-2, -1, 0, 1, 2	5	10
f	-3, -2, -1, 0, 1, 2, 3	7	14

Table 3.4 Occupancy of subshells

Because of the $n + l$ rule, the shells do not all fill before the previous shells have been completed.

To write the detailed electron configuration of any atom, showing how many electrons occupy each of the various subshells, one needs to know only the order of increasing energy of the subshells and the maximum number of electrons that will fit into each (given in Table 3.4). A convenient way to designate such a configuration is to write the shell and subshell designation and add a superscript to denote the number of electrons occupying that subshell. For example, the electronic configuration of the sodium atom is written as follows:

$$\text{Na} \qquad 1s^2 2s^2 2p^6 3s^1$$

The 3*s* subshell can hold a maximum of two electrons, but in this atom this subshell is not filled. The total number of electrons in the atom can easily be determined by adding the numbers in all the subshells, that is, by adding all the superscripts. For sodium, this sum is 11, equal to the atomic number of sodium.

Orbital Shape

The **Heisenberg uncertainty principle** requires that, since the energy of the electron is known, its exact position cannot be known. It is possible to learn only the probable location of the electron in the vicinity of the atomic nucleus. An approximate description may be given in terms of values of the quantum numbers n, l, and m. The shapes of the first few orbitals are shown in Figure 3-1 for the case of the hydrogen atom. This figure shows that, in general, an electron in the 1*s* orbital is equally likely to be found in any direction about the nucleus. The maximum probablility is at a distance corresponding to the experimentally determined radius of the hydrogen atom. In contrast, in the case of an electron in a 2*p* orbital, there are three possible values of the quantum number m. There are three possible regions in which the electron is most likely to be found. It is customary to depict these orbitals as being located along the cartesian (x, y, and z) axes of a three dimensional graph. Hence, the three probability distributions are labeled p_x, p_y, and p_z, respectively.

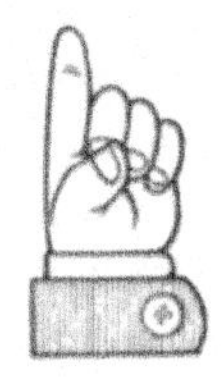

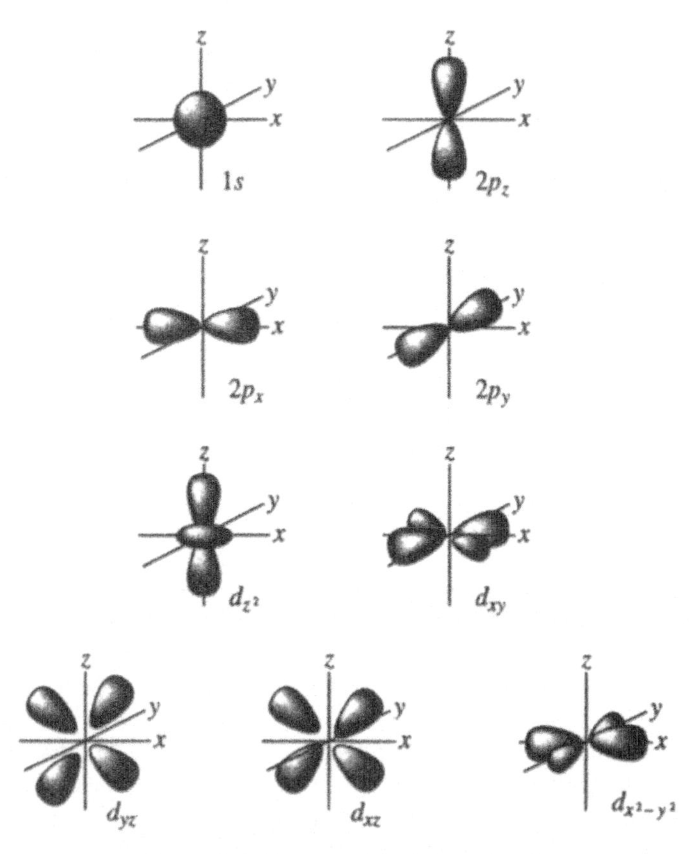

Figure 3-1 Shapes of various orbitals

Buildup Principle

With each successive increase in atomic number, a given atom has one more electron than the previous atom. Thus, it is possible to start with hydrogen and, adding one electron at a time, build up the electronic configuration of each of the other elements. In the buildup of electronic structures of the atoms of the elements, the last electron is added to the lowest-energy subshell possible. The relative order of the energies of all the important subshells in an atom is shown in Figure 3-2. The energies of the various subshells are plotted along the vertical axis. The subshells

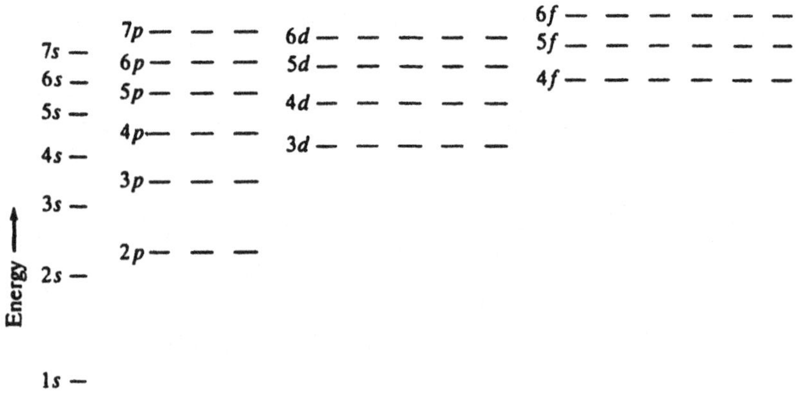

Energy

7s, 6s, 5s, 4s, 3s, 2s, 1s
7p, 6p, 5p, 4p, 3p, 2p
6d, 5d, 4d, 3d
6f, 5f, 4f

Figure 3-2 Energy level diagram

are displaced left to right merely to avoid overcrowding. The order of increasing energy is as follows:

1*s*, 2*s*, 2*p*, 3*s*, 3*p*, 4*s*, 3*d*, 4*p*, 5*s*, 4*d*, 5*p*, 6*s*, 4*f*, 5*d*, 6*p*, 7*s*, 5*f*, 6*d*.

Diagrams such as the one in Figure 3-2 may be used to draw electronic configurations (see Figure 3-3). Arrows pointing up are used to represent electrons with one given spin, and arrows pointing down are used to represent electrons with the opposite spin. In atoms with partially filled *p*, *d*, or *f* subshells, the electrons stay unpaired as much as possible. This effect is called **Hund's rule of maximum multiplicity**.

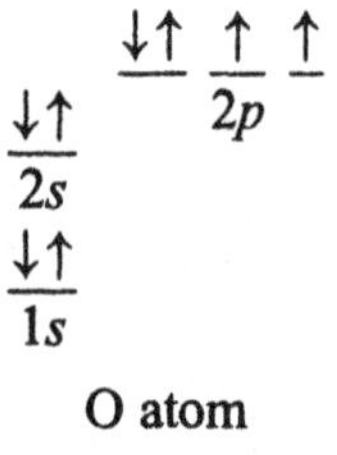

↓↑ ↑ ↑
2*p*
↓↑
2*s*
↓↑
1*s*

O atom

Figure 3-3 Electron configuration of an oxygen atom

Be able to distinguish between Pauli's exclusion principle, the Heisenberg uncertainty principle, and Hund's rule of maximum multiplicity.

It turns out that fully filled or half-filled subshells have greater stability than subshells having some other numbers of electrons. One effect of this added stability is the fact that some elements do not follow the $n + l$ rule exactly. For example, copper would be expected to have a configuration

$n + l$ configuration for Cu	$1s^2 2s^2 2p^6 3s^2 3p^6 4s^2 3d^9$
Actual configuration for Cu	$1s^2 2s^2 2p^6 3s^2 3p^6 4s^1 3d^{10}$

The actual configuration has two subshells of enhanced stability ($3d$ and $4s$) in contrast to one subshell ($4s$) of the expected configuration. (There are also some elements whose configurations do not follow the $n + l$ rule and which are not enhanced by the added stability of an extra fully filled and half-filled subshells.)

Electronic Structure and the Periodic Table

The arrangement of electrons in successive energy levels in the atom provides an explanation of the periodicity of the elements, as found in the periodic table. The charges on the nuclei of the atoms increase in a regular manner as the atomic number increases. Therefore, the number of electrons surrounding the nucleus increases also. The number and arrangement of the electrons in the outermost shell of an atom vary in a periodic manner. For example, all the elements in Group IA (H, Li, Na, K, Rb, Cs, Fr), corresponding to the elements that begin a new row or period, have electronic configurations with a single electron in the outermost shell, specifically, an s subshell.

The noble gases, located at the end of each period, have electronic configurations of the type ns^2np^6, where n represents the number of the outermost shell. Also, n is the number of the period in the periodic table in which the element is found.

Since atoms of all elements in a given group of the periodic table have analogous arrangements of electrons in their outermost shells and different arrangements from elements of other groups, it is reasonable to conclude that the outermost electronic configuration of the atom is responsible for the chemical characteristics of the element. Elements with similar arrangements of electrons in their outer shells will have similar properties. For example, the formulas of their oxides will be of the same type. The electrons in the outermost shells of the atoms are referred to as **valence electrons.**

As the atomic numbers of the elements increase, the arrangements of electrons in successive energy levels vary in a periodic manner. As shown in Figure 3-2, the energy of the 4*s* subshell is lower than that of the 3*d* subshell. Therefore, at atomic number 19, corresponding to the element potassium, the 19th electron is found in the 4*s* subshell rather than the 3*d* subshell. The fourth shell is started before the third shell is completely filled. At atomic number 20, calcium, a second electron completes the 4*s* subshell. Beginning with atomic number 21 and continuing through the next nine elements, successive electrons enter the 3*d* subshell. When the 3*d* subshell is complete, the following electrons occupy the 4*p* subshell through atomic number 36, krypton. In other words, for elements 21 through 30, the last electrons added are found in the 3*d* subshell rather than the valence shell. The elements Sc through Zn are called **transition elements**, or *d* block elements. A second series of transition elements begins with yttrium, atomic number 39, and includes ten elements. This series corresponds to the placement of ten electrons in the 4*d* subshell.

The elements maybe divided into types (see Figure 3-4), according to the position of the last electron added to those present in the preceding element. In the first type, the last electron added enters the valence shell. These elements are called **main group elements**. In the second type, the last electron enters a *d* subshell in the next-to-last shell. These elements are called **transition elements**. The third type has the last electron enter the *f* subshell in the $n - 2$ shell, the second shell below the valence shell. These elements are called the **inner transition elements**.

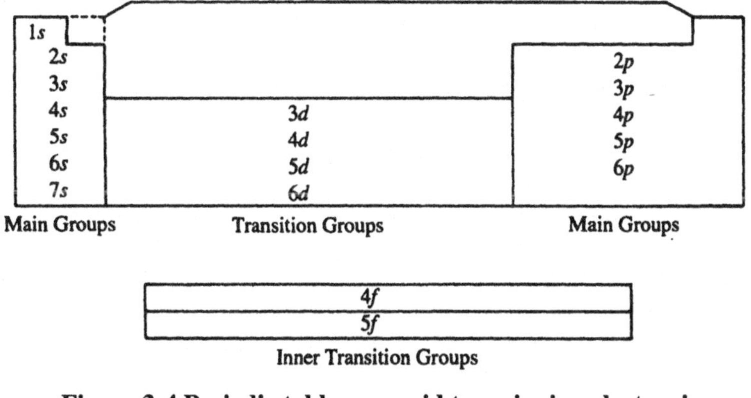

Figure 3-4 Periodic table as an aid to assigning electronic configurations

An effective way to determine the detailed electronic configuration of any element is to use the periodic table to determine which subshell to fill next. Each *s* subshell holds a maximum of two electrons; each *p* subshell holds a maximum of six electrons; each *d* subshell holds a maximum of ten electrons; and each *f* subshell holds a maximum of 14 electrons. These numbers match the numbers of elements in a given period in the various blocks. To get the electronic configuration, start at hydrogen (atomic number = 1) and continue in order of atomic number using the periodic table.

Instead of writing out the entire electronic configuration of an atom, especially an atom with many electrons, we sometimes abbreviate the configuration by using the configuration of the previous noble gas and represent the rest of the electrons explicitly. For example, Ni may be abbreviated as

$$[\mathrm{Ar}]4s^2 3d^8$$

Solved Problems

Solved Problem 3.1 What is the maximum number of electrons that can occupy the M shell?

Solution: The M shell in an atom corresponds to the third energy level ($n = 3$); hence the maximum number of electrons it can hold is

$$2n^2 = 2(3)^2 = 2 \times 9 = 18$$

Solved Problem 3.2 What values are permitted for (*a*) l if $n = 4$; (*b*) m if $l = 3$; and (*c*) s if $n = 4$, $l = 1$, and $m = {}^-1$?

Solution: (*a*) Since l can have integer values from 0 up to $n - 1$, values of 0, 1, 2, and 3 are permitted; (*b*) Since m can have integer values from ^-l through 0 up to ^+l, values $^-3$, $^-2$, $^-1$, 0, 1, 2, and 3 are permitted; (*c*) The value for s is independent of any of the other quantum numbers. It must be either $^-½$ or $^+½$.

Solved Problem 3.3 Arrange the electrons in the following list in order of increasing energy, lowest first:

	n	l	m	s
(*a*)	3	2	$^-1$	½
(*b*)	4	0	0	$^-½$
(*c*)	4	1	1	$^-½$
(*d*)	2	1	$^-1$	½

Solution: Electron (*d*) has the lowest value of $n + l$ $(2 + 1 = 3)$, and so it is lowest in energy of the four electrons. Electron (*b*) has the next lowest sum of $n + l$ $(4 + 0 = 4)$ and is next in energy (despite the fact that it does not have the next lowest n value). Electrons (*a*) and (*c*) both have the same sum of $n + l$ $(3 + 2 = 4 + 1 = 5)$. Therefore, in this case, electron (*a*), the one with the lower n value, is lower in energy. Electron (*c*) is highest in energy.

Solved Problem 3.4 Write the electronic configuration of Te, atomic number 52.

Solution: $1s^22s^22p^63s^23p^64s^23d^{10}4p^65s^24d^{10}5p^4$.

Chapter 4
CHEMICAL BONDING

IN THIS CHAPTER:

- ✔ *Chemical Formulas*
- ✔ *The Octet Rule*
- ✔ *Ions*
- ✔ *Electron Dot Notation*
- ✔ *Covalent Bonding*
- ✔ *Ionic vs. Covalent Bonds*
- ✔ *Solved Problems*

Chemical Formulas

Most materials found in nature are compounds or mixtures of compounds rather than free elements. It is a rule of nature that the state which is most probably encountered corresponds to the state of lowest energy. Since compounds are encountered more often than free elements, it can be inferred that the combined state must be the state of lower energy. Indeed, those elements that do occur naturally as free elements must posses some characteristics that correspond to a relatively low energy state. The chemical combination corresponds to the tendency of atoms to assume the most stable electronic configuration possible. Before studying the forces holding the particles

together in a compound, however, we must first understand the meaning of a **chemical formula**.

Writing a formula implies that the atoms in the formula are bonded together in some way. The relative numbers of the atoms of the elements in a compound are shown in a chemical formula by writing the symbols of the elements followed by appropriate subscripts to denote how many atoms of each element there are in the formula unit. A subscript following the symbol gives the number of atoms of that element per formula unit. If there is no subscript, one atom per formula unit is implied. For example, the formula H_2SO_4 describes a molecule containing two atoms of hydrogen, one atom of sulfur, and four atoms of oxygen. Sometimes groups of atoms which are bonded together within a molecule or ionic compound are grouped in the formula within parentheses. The number of such groups is indicated by a subscript following the closing parenthesis. For example, the 3 in $(NH_4)_3PO_4$ states that there are three NH_4 groups present per formula unit. There is only one PO_4 group; therefore parentheses are not around it.

Remember

Formulas give the ratio of atoms of each element to all the others in a compound.

Note that a pair of hydrogen atoms bonded together is a hydrogen molecule. Seven elements, when uncombined with other elements, form **diatomic** (having two atoms) molecules. These elements are hydrogen, nitrogen, oxygen, fluorine, chlorine, bromine, and iodine. They are easy to remember because the last six form the shape of a large "7" in the periodic table, starting at element 7, nitrogen.

The Octet Rule

The elements helium, neon, argon, krypton, xenon, and radon, known as the noble gases, occur in nature as **monatomic** (having only one atom) molecules. In nature, their atoms are not combined with atoms of other elements or with other atoms like themselves. These elements are very stable due to the electronic structures of their atoms. The charge on the nucleus and the number of electrons in the valence shell determine the chemical properties of the atom. The electronic configurations of the noble gases (except for that of helium) correspond to a valence shell containing eight electrons, a very stable configuration called an **octet**. Atoms of other main group elements tend to react with other atoms in various ways to achieve the octet. The tendency to achieve an octet of electrons in the outermost shell is called the **octet rule**. If there is only one shell occupied, then the maximum number of electrons is two. A configuration of two electrons in the first shell, with no other shells occupied by electrons, is stable and therefore is also said to obey the octet rule.

Ions

The electronic configuration of a sodium atom is

$$\text{Na} \quad 2 \quad 8 \quad 1 \quad (1s^2 2s^2 2p^6 3s^1)$$

If sodium were to lose one electron, the resulting species would have the configuration

$$\text{Na}^+ \quad 2 \quad 8 \quad (1s^2 2s^2 2p^6)$$

The nucleus of a sodium atom contains 11 protons, and if there are only ten electrons surrounding the nucleus, the atom will have a net charge of $^+1$. An atom (or group of atoms) that contains a net charge is called an **ion**. In chemical notation, an ion is represented by the symbol of the atom with the charge indicated as a superscript to the right. Thus, the sodium ion is written Na^+. The sodium ion has the same configuration of electrons as a neon atom (atomic number 10). Ions that have the electronic configurations of noble gases are rather stable. Note the very important differences between a sodium ion and a neon atom: the different nuclear charges and the net $^+1$ charge on Na^+. The Na^+ ion is not as stable as the Ne atom.

Compounds, even ionic compounds, have no net charge. In the compound sodium chloride, there are sodium ions (Na^+) and chloride ions (Cl^-); the oppositely charged ions attract one another and form a regular geometric arrangement, as shown in Figure 4-1. This attraction is called an **ionic bond**. There are equal numbers of Na^+ and Cl^- ions, and the compound is electrically neutral. It would be inaccurate to speak of a molecule of solid sodium chloride or of a bond between a specific sodium ion

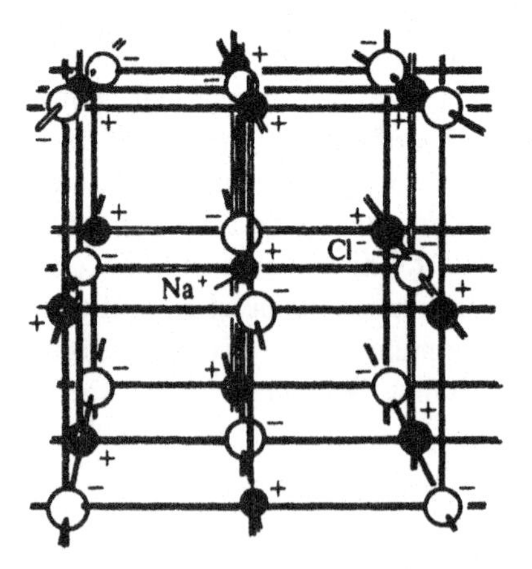

Figure 4-1 Ball-and-stick model of the sodium chloride structure

and a specific chloride ion. The substance NaCl is extremely stable because of: (1) the stable electronic configurations of the ions and (2) the attractions between the oppositely charged ions.

The electronic configurations of ions of many main group elements and even a few transition elements can be predicted by assuming that the gain or loss of electrons by an atom results in a configuration analogous to that of a noble gas, which contains an octet of electrons in the valence shell. Not all the ions that could be predicted with this rule actually form. For example, few monatomic ions have charges of $^+4$, and none have charges of $^-4$. Positively charged ions are called **cations**, and negatively charged ions are called **anions**.

Electron Dot Notation

Electron dot notation is often used to represent the formation of bonds between atoms. In this notation, the symbol for an element represents the nucleus of an atom of the element plus all the electrons except those in the outermost (valence) shell. The outermost electrons are represented by dots. For example, the dot notation for the first ten elements in the periodic table is as follows:

H· He: Li· Be: :B·

:Ċ· :N̤· :Ö̤: :F̤̈: :N̤̈e:

Using electron dot notation, the production of calcium oxide may be pictured. The calcium atom has two electrons in its outermost shell. Each oxygen atom has six electrons in its outermost shell and requires two more electrons to attain its octet. Each oxygen atom therefore requires one calcium atom from which to obtain the two electrons, and calcium and oxygen react in a 1:1 ratio.

$$\text{Ca:} + \text{:}\dot{\text{O}}\text{:} \longrightarrow \text{Ca}^{2+} + \text{:}\ddot{\text{O}}\text{:}^{2-}$$

Covalent Bonding

The element hydrogen exists in the form of diatomic molecules, H_2. Since both hydrogen atoms are identical, they are not likely to have opposite charges. Each free hydrogen atom contains a single electron, and if the atoms are to achieve the same electronic configuration as atoms of helium, they must each acquire a second electron. If two atoms of hydrogen are allowed to come sufficiently close to each other, their two electrons will effectively belong to both atoms. The positively charged hydrogen nuclei are attracted to the pair of electrons shared between them, and effectively, a bond is formed. The bond formed from the sharing of a pair of electrons (or more than one pair) between two atoms is called a **covalent bond**. The diatomic hydrogen molecule is more stable than two separate hydrogen atoms. Other pairs of nonmetallic atoms share electrons in the same way.

The formation of covalent bonds between atoms can be depicted by electron dot notation. The formation of some covalent bonds is shown in this manner below:

$$\text{H}\cdot + \cdot\text{H} \longrightarrow \text{H:H}$$

$$\text{:}\ddot{\text{I}}\cdot + \cdot\ddot{\text{I}}\text{:} \longrightarrow \text{:}\ddot{\text{I}}\text{:}\ddot{\text{I}}\text{:}$$

$$\text{H}\cdot + \cdot\ddot{\text{I}}\text{:} \longrightarrow \text{H:}\ddot{\text{I}}\text{:}$$

$$\cdot\dot{\text{C}}\cdot + 4\,\cdot\ddot{\text{I}}\text{:} \longrightarrow \begin{array}{c} \text{:}\ddot{\text{I}}\text{:} \\ \text{:}\ddot{\text{I}}\text{:C:}\ddot{\text{I}}\text{:} \\ \text{:}\ddot{\text{I}}\text{:} \end{array}$$

In these examples, it can be seen that the carbon and iodine atoms can achieve octets of electrons by sharing pairs of electrons with other atoms.

Hydrogen atoms attain **duets** of electrons because the first shell is complete when it contains two electrons.

Every group of electrons shared between two atoms constitutes a covalent bond. When one pair of electrons is involved, the bond is called a **single bond**. Sometimes it is necessary for two atoms to share more than one pair of electrons to attain octets. When two pairs of electrons unite two atoms, the bond is called a **double bond**. Three pairs of electrons shared between two atoms constitute a **triple bond**. Examples of these types of bonds are given below:

H:H	:Ö::C::Ö:
Single bond	Two double bonds
:N:::N:	:F̤̈:S̤̈:F̤̈:
Triple bond	Two single bonds

The constituent atoms in polyatomic ions are also linked by covalent bonds. In these cases, the net charge on the ion is determined by the total number of electrons and the total number of protons. For example, the ammonium ion NH_4^+, formed from five atoms, contains one fewer electron (10) than the number of protons (11).

To write electron dot diagrams for molecules that contain several atoms, first determine the number of valence electrons available in each atom. Second, determine the number of electrons necessary to satisfy the octet rule with no sharing. The difference between the numbers in these first two steps is the number of bonding electrons. Place the atoms as symmetrically as possible. Place the number of electrons to be shared between the atoms, one pair at a time, at first one pair between each pair of atoms. Use as many pairs as remain to make double or triple bonds. Add the remainder of available electrons to complete the octets of all the atoms. There should be just enough if the molecule or ion follows the octet rule. (There are exceptions, which will not be covered here.)

You Need to Know ✓

In polyatomic ions, more or fewer electrons are available than the number that comes from the valence shells of the atoms in the ion. This gain or loss of electrons provides the charge.

Ionic vs. Covalent Bonds

The word **bonding** applies to any situation in which two or more atoms are held together in such close proximity that they form a characteristic species that has distinct properties and that can be represented by a chemical formula. In compounds consisting of ions, bonding results from attractions between the oppositely charged ions. In such compounds in the solid state, each ion is surrounded on all sides by ions of the opposite charge. In a solid ionic compound, it is incorrect to speak of a bond between specific pairs of ions. In contrast, covalent bonding involves the sharing of electron pairs between two specific atoms, and it is possible to speak of a definite bond.

Polyatomic ions possess covalent bonds as well as an overall charge. The charges on polyatomic ions cause ionic bonding between these groups of atoms and oppositely charged ions. In writing electron dot structures, the distinction between ionic and covalent bonds must be clearly indicated. For example, an electron dot diagram for the compound NH_4NO_3 would be

$$\left[\begin{array}{c} H \\ H:\ddot{N}:H \\ H \end{array}\right]^{+} \qquad \left[\begin{array}{c} :\ddot{O}::\underset{..}{N}:\ddot{O}: \\ :\underset{..}{O}: \end{array}\right]^{-}$$

Electronegativity is a semiquantitative measure of the ability of an atom to attract electrons involved in covalent bonds. Atoms with higher electronegativities have greater electron-attracting ability. Some electronegativity values are given in Figure 4-2. The greater the electroneg-

H 2.1							He
Li 1.0	Be 1.5	B 2.0	C 2.5	N 3.0	O 3.5	F 4.0	Ne
Na 0.9	Mg 1.2	Al 1.5	Si 1.8	P 2.1	S 2.5	Cl 3.0	Ar
K 0.8	Ca 1.0	Ga 1.6	Ge 1.8	As 2.0	Se 2.4	Br 2.8	Kr
Rb 0.8	Sr 1.0	In 1.7	Sn 1.8	Sb 1.9	Te 2.1	I 2.5	Xe
Cs 0.7	Ba 0.9	Tl 1.8	Pb 1.9	Bi 1.9	Po 2.0	At 2.2	Rn

Figure 4-2 Selected electronegativities

ativity difference between a pair of elements, the more likely they are to form an ionic compound; the lower the difference in electronegativity, the more likely the compound will be covalent.

Most **binary compounds** (made of two elements) of metals with nonmetals are essentially ionic. All compounds involving only nonmetals are essentially covalent except for compounds containing the NH_4^+ ion. Practically all **tertiary compounds** (made of three elements) contain covalent bonds. If one or more of the elements is a metal, there is likely to be ionic as well as covalent bonding in the compound.

When some molecules containing only covalent bonds are dissolved in water, the molecules react with the water to produce ions in solution. For example, HCl contains a covalent bond. However, when HCl is dissolved in water, H_3O^+ (**hydronium ion**, abbreviated H^+) and Cl^- ions form.

Solved Problems

Solved Problem 4.1 How many H atoms and how many S atoms are there per formula unit in $(NH_4)_2SO_4$?

Solution: There are two NH_4 groups, each containing four H atoms, for a total of eight H atoms per formula unit. There is only one S atom; the 4 defines the number of O atoms.

Solved Problem 4.2 Predict the charge on a calcium ion and that on a bromide ion, and deduce the formula of calcium bromide.

Solution: The electronic configuration of a calcium atom is

Ca 2 8 8 2 $(1s^2 2s^2 2p^6 3s^2 3p^6 4s^2)$

By losing two electrons, calcium attains the electronic configuration of argon and thereby acquires a charge of $^+2$.
A bromine atom has the configuration

Br 2 8 18 7 $(1s^2 2s^2 2p^6 3s^2 3p^6 4s^2 3d^{10} 4p^5)$

By gaining one electron, the bromine atom attains the electronic configuration of krypton and also attains a charge of $^-1$. The two ions expected are therefore Ca^{2+} and Br^-. Since calcium bromide as a whole cannot have any net charge, there must be two bromide ions for each calcium ion; hence, the formula is $CaBr_2$.

Solved Problem 4.3 Draw an electron dot diagram for (*a*) Na_2SO_4 and (*b*) H_2SO_4. What is the major difference between them?

Solution:

Ans. (*a*) $2\,Na^+$ [:Ö: / :Ö:S:Ö: / :O:]$^{2-}$ (*b*) :Ö: / H:Ö:S:Ö:H / :O:

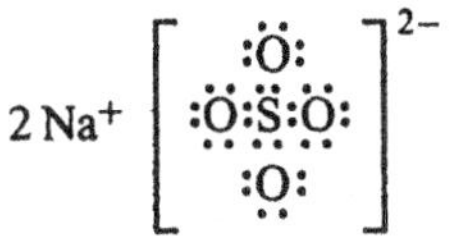

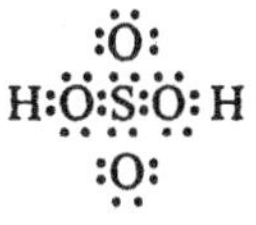

In the sodium salt there is ionic bonding as well as covalent bonding; in the hydrogen compound, there is only covalent bonding.

Solved Problem 4.4 Which element of each of the following pairs has the higher electronegativity? (*a*) Mg and Cl, (*b*) S and O, and (*c*) Cl and O.

Solution: (*a*) Cl. It lies farther to the right in the periodic table. (*b*) O. It lies farther up in the periodic table. (*c*) O. It lies farther up in the periodic table, and is more electronegative even though it lies one group to the left (see Figure 4-2).

Solved Problem 4.5 What is the electronic configuration of Fe^{2+}?

Solution: First write the electronic configuration of the neutral atom, then remove two electrons from the subshell with the highest principal quantum number, in this case the two 4*s* electrons. They are the electrons in the outermost shell, despite that fact that they were not the last to be added. The configuration is

$$Fe^{2+} \quad 1s^22s^22p^63s^23p^64s^03d^6 \text{ or } 1s^22s^22p^63s^23p^63d^6$$

Chapter 5

Inorganic Nomenclature

In This Chapter:

- ✔ *Binary Compounds of Nonmetals*
- ✔ *Ionic Compounds*
- ✔ *Inorganic Acids*
- ✔ *Acid Salts*
- ✔ *Hydrates*
- ✔ *Solved Problems*

Binary Compounds of Nonmetals

Binary compounds with two nonmetals are named with the element to the left or below in the periodic table named first. The other element is then named, with its ending changed to –**ide** and a prefix added to denote the number of atoms of that element present. If there is more than one atom of the first element present, a prefix is used with the first element also. If one of the elements is to the left and the other below, the one to the left is named first unless that element is oxygen or fluorine, in which case it is named last. The same order of elements is used in writing formulas for these compounds. The element with the lower electronegativity is usually named first. The prefixes are listed in Table 5.1.

Number of Atoms	Prefix
1	mono– or mon–
2	di–
3	tri–
4	tetra– or tetr–
5	penta– or pent–
6	hexa–
7	hepta–
8	octa–
9	nona–
10	deca–

Table 5.1 Prefixes for nonmetal-nonmetal compounds

The systematic names presented for binary nonmetal-nonmetal compounds are not used for hydrogen compounds of group III, IV, and V elements or for water. These compounds have common names which are used instead. Water (H_2O) and ammonia (NH_3) are the most important compounds in this class.

Ionic Compounds

Ionic compounds are composed of cations and anions. The cation is always named first. Naming of the cation depends on whether the ion is monatomic. If not, special names are given, such as ammonium for NH_4^+ and mercury(I) ion or mercurous ion for Hg_2^{2+}. If the cation is monatomic, the name depends on whether the element forms more than one positive ion in its compound. For example, sodium forms only one positive ion in all its compounds, Na^+. Iron forms two positive ions, Fe^{2+} and Fe^{3+}. Cations of elements that form only one type of ion in all their compounds need not be further identified in the name. Cations of metals that occur with two or more different charges must be further identified. For monatomic cations, we use a Roman numeral in parentheses attached to the name to indicate the charge on such ions. Thus, Fe^{2+} is called the iron(II) ion, and Fe^{3+} is called the iron(III) ion.

The elements that form only one cation are the alkali metals (group IA), the alkaline earth metals (group IIA), zinc, cadmium, aluminum, and most often silver. The charge on the ions that these elements form in their compounds is always equal to their classical periodic table group number.

Don't Get Confused!

Roman numerals in the names refer to the charge on the ion. Arabic numeral subscripts in the formula refer to the number of atoms.

An older system for naming cations of elements having more than one possible cation uses the ending **–ic** for the ion with the higher charge and the ending **–ous** for the ion with the lower charge.

Common anions may be grouped as follows: monatomic anions, oxyanions, and special anions. There are special endings for the first two groups; the third group is small enough to be memorized.

If the anion is monatomic, the name of the element is amended by changing the ending to –ide. Note that this is the same ending used for binary nonmetal-nonmetal compounds. All monatomic anions have names ending in –ide, but there are few anions that consist of more than one atom which also end in –ide, the most important of these are the hydroxide ion (OH^-) and the cyanide ion (CN^-).

Oxyanions consist of an atom of an element plus some number of atoms of oxygen covalently bonded to it. The name of the anion is given by the name of the element with its ending changed to either **–ate** or **–ite**. In some cases, it is also necessary to add the prefix **per–** or **hypo–** to distinguish all the possible oxyanions from one another. For example, there are four oxyanions of chlorine, which are named as follows:

ClO_4^- perchlorate ion
ClO_3^- chlorate ion
ClO_2^- chlorite ion
ClO^- hypochlorite ion

One may think of the –ite ending as meaning "one fewer oxygen atom." The per– and hypo– prefixes then mean "one more oxygen atom" and "still one fewer oxygen atom," respectively. Other elements have similar sets of oxyanions, but not all have four different oxyanions.

You Need to Know ✓

These oxyanions:

Chlorate	ClO_3^-
Bromate	BrO_3^-
Iodate	IO_3^-
Nitrate	NO_3^-
Phosphate	PO_4^{3-}
Sulfate	SO_4^{2-}
Carbonate	CO_3^{2-}

There are a few special anions that seem rather unusual but are often used in general chemistry (see Table 5.2).

CrO_4^{2-}	Chromate
$Cr_2O_7^{2-}$	Dichromate
MnO_4^-	Permanganate
$C_2H_3O_2^-$	Acetate
CN^-	Cyanide
OH^-	Hydroxide
O_2^{2-}	Peroxide

Table 5.2 Special anions

To name an ionic compound, simply put the names of the cation and anion together, in that order. The number of cations and anions per formula need not be included in the compound name because anions have characteristic charges, and the charge of the cation has already been established by its name. There are as many cations and anions as needed to get a neutral compound with the lowest possible integral subscripts.

Inorganic Acids

The anions described in the preceding section may be formed by reaction of the corresponding acids with hydroxides For example,

$$HCl + NaOH \rightarrow NaCl + H_2O$$

The salts formed by these reactions consist of cations and anions. The cation in the above example is Na^+ and the anion is Cl^-. The anion in this reaction is formed from its parent acid. Thus, the acid and the anion are related, and so are their names.

When they are pure, acids are not ionic. When we put them into water solution, seven become fully ionized: HCl, $HClO_3$, $HClO_4$, HBr, HI, HNO_3, and H_2SO_4. They are called **strong acids**. All other acids ionize at least to some extent and are called **weak acids**. Both types react completely with hydroxides to form ions. Formulas for acids conventionally are written with the hydrogen atoms which can ionize first. Different names for some acids are given when the compound is pure and when it is dissolved in water. For example, HCl is called hydrogen chloride when it is in the gas phase, but in water it ionizes to give hydrogen ions and chloride ions and is called hydrochloric acid. The names for all the acids corresponding to the anions in Table 5.2 and their derivatives can be deduced by the following simple rules:

1. Replace the –ate ending of an anion with "–ic acid" or replace the –ite ending with "–ous acid."

2. If the anion ends in –ide, add the prefix hydro- and change the ending to "–ic acid."

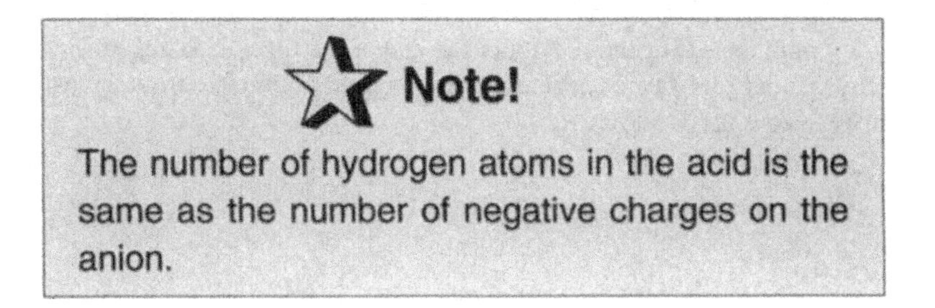

Acid Salts

It is possible for an acid with more than one ionizable hydrogen atom to react with fewer hydroxide ions and to form a product with some ionizable hydrogen atoms left. For example,

$$H_2SO_4 + NaOH \rightarrow NaHSO_4 + H_2O \; (Na^+ \text{ and } HSO_4^-)$$

The products are called **acid salts**, and each anion contains at least one ionizable hydrogen atom and at least one negative charge. The sum of the negative charges plus hydrogen atoms equals the original number of hydrogen atoms in the parent acid and also the number of negative charges in the normal anion. For example, HSO_4^- contains one hydrogen atom plus one negative charge, for a total of two. That is the number of hydrogen atoms in H_2SO_4 and also the number of negative charges in SO_4^{2-}.

The anions of acid salt are named with the word "hydrogen" placed before the name of the normal anion. Thus, HSO_4^- is the hydrogen sulfate ion. To denote two atoms, the prefix **di**– is used. HPO_4^{2-} is the hydrogen phosphate ion, while $H_2PO_4^-$ is the dihydrogen phosphate ion. In an older naming system, the prefix **bi**– was used instead of the word hydrogen when one of two hydrogen atoms was replaced. Thus, HCO_3^- was called the bicarbonate ion instead of the more modern name, hydrogen carbonate ion.

Hydrates

Some stable ionic compounds are capable of bonding to a certain number of molecules of water per formula unit. Thus, copper(II) sulfate forms the stable $CuSO_4 \cdot 5H_2O$, with five molecules of water per $CuSO_4$ unit. This type of compound is called a **hydrate**. The name of the compound is the name of the **anhydrous** (without water) compound with a designation for the number of water molecules appended. Thus, $CuSO_4 \cdot 5H_2O$ is called copper(II) sulfate pentahydrate. The 5 multiples everything after it until the next centered dot or the end of the formula. Thus, included in $CuSO_4 \cdot 5H_2O$ are ten H atoms and nine O atoms (five from the water and four in the sulfate ion).

Solved Problems

Solved Problem 5.1 Name P_2O_5.

Solution: Phosphorus lies to the left and below oxygen in the periodic table, so it is named first: diphosphorus pentoxide.

Solved Problem 5.2 Write the formula for sulfur hexafluoride.

Solution: SF_6. Sulfur is named first, since it lies below and to the left of fluorine in the perioidic table. The prefixes tell how many atoms of the second element is in each molecule.

Solved Problem 5.3 Write the formulas for (*a*) copper(I) oxide and (*b*) copper(II) oxide.

Solution: (*a*) Cu_2O. The copper(I) has a charge of $^{+}1$, and therefore two copper(I) ions are required to balance the $^{-}2$ charge on one oxide ion. (*b*) CuO. The copper(II) ion has a charge of $^{+}2$, and therefore one such ion is sufficient to balance the $^{-}2$ charge on the oxide ion.

Solved Problem 5.4 Name the following ions: (*a*) SO_3^{2-} and (*b*) IO_4^{-}.

Solution: (a) Remembering that sulfate is SO_4^{2-}, we note that this ion has one fewer oxygen atom. It is the sulfite ion. (*b*) Remembering that iodate is IO_3^{-}, we note that this ion has one more oxygen atom. It is the periodate ion.

Solved Problem 5.5 Name $Ba(NO_3)_2$.

Solution: The cation is the barium ion. The anion is the nitrate ion. The compound is barium nitrate. It is not necessary to state anything to indicate the presence of two nitrate anions; that can be deduced from the fact that the barium ion has a $^{+}2$ charge and nitrate has a $^{-}1$ charge.

Solved Problem 5.6 Name the following acids: (*a*) HI, (*b*) HNO_3, and (*c*) $HClO_4$.

Solution: (*a*) HI is related to I^-, the iodide ion. The –ide ending is changed to –ic acid and the prefix hydro– is added. The name is hydroiodic acid. (*b*) HNO_3 is related to NO_3^-, the nitrate ion. The –ate ending is changed to –ic acid. The name is nitric acid. (*c*) $HClO_4$ is related to ClO_4^-, the perchlorate ion. The prefix per– is not changed, but the ending is changed to –ic acid. The name is perchloric acid.

Solved Problem 5.7 What is the formula for the dihydrogen phosphate ion?

Solution: $H_2PO_3^-$. Note that the two hydrogen atoms plus the one charge total three, equal to the number of hydrogen atoms in phosphoric acid.

Solved Problem 5.8 Name the following compound, and state how many hydrogen atoms it contains per formula unit.

$$Na_2SO_3{\cdot}7H_2O$$

Solution: Sodium sulfite heptahydrate. It contains 14 H atoms per formula unit.

Chapter 6 FORMULA CALCULATIONS

IN THIS CHAPTER:

- ✔ *Formula Units and Masses*
- ✔ *The Mole*
- ✔ *Percent Compositions of Compounds*
- ✔ *Empirical Formulas*
- ✔ *Molecular Formulas*
- ✔ *Solved Problems*

Formula Units and Masses

Some elements combine by covalent bonding into molecules. Other elements combine by gaining or losing electrons to form ions. The ions are attracted to one another, a type of bonding called ionic bonding, and form combinations containing millions or more of ions of each element. To identify these ionic compounds, the simplest formula is generally used. Formulas for ionic compounds such as NaCl represent one **formula unit**. However, formula units may also refer to uncombined atoms (such as Au) or molecules (such as CH_4).

The **formula mass** of a compound is the sum of the atomic masses of all the atoms in the formula. Thus, in the same way a symbol is used to represent an element, a formula is used to represent a compound and also one unit of the compound. The formula mass of a compound is easily determined on the basis of the formula. Note that just as a formula unit may refer to uncombined atoms, molecules, or atoms combined in an ionic compound, the term formula mass may refer to the atomic mass of an atom, the molecular mass of a molecule, or the formula mass of a formula unit of an ionic compound.

The Mole

Atoms and molecules are incredibly small. Just as the dozen is used as a convenient number of items in everyday life, chemists use the **mole** to describe quantities of atoms or molecules. One mole is 6.02×10^{23} items, a number called **Avogadro's number**. Using moles simplifies calculations. The number of formula units can be converted to moles of the same substance, and vice versa, using Avogadro's number. Some authors refer to a mole as a "gram molecular mass" because one mole of molecules has a mass in grams equal to its molecular mass. Mole is abbreviated **mol**.

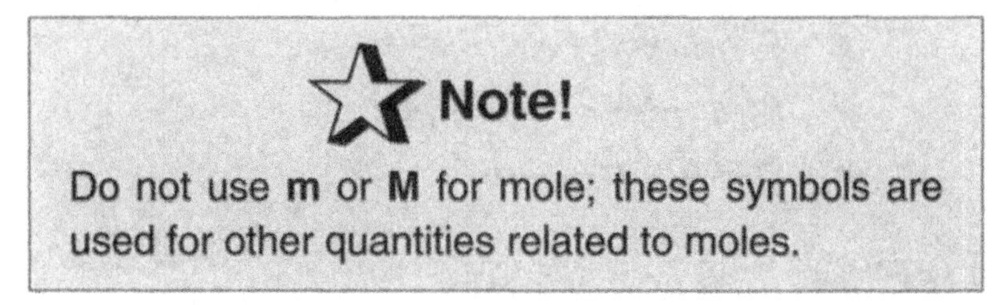

The formula mass of a substance is equal to its number of grams per mole. Avogadro's number is the number of atomic mass units in one g. It is defined in that manner so that the atomic mass of an element (in amu) is numerically equal to the number of grams of the element per mole. We use the term **molar mass** for the mass of one mol of any substance. The units are typically grams per mole. For example, the mass of 1.00 mol K is 39.1 g.

The number of moles of each element in a mole of a compound is stated in the chemical formula. Hence, the formula can be used to convert the number of moles of the compound to the number of moles of its component elements, and vice versa.

Percent Compositions of Compounds

The term **percent** means the quantity or number of units out of 100 units total. Percentage is computed by finding the fraction of the total quantity represented by the quantity under discussion, and multiplying by 100 percent. The concept of percentage is often used to describe the composition of compounds. If the formula of a compound is known, the percent mass of an element in the compound is determined by computing the fraction of the formula mass that is made up of that element and multiplying that fraction by 100 percent. The formula gives the number of moles of atoms of each element in each mole of the compound; then the number of grams of each element in the number of grams of compound in one mol of the compound is fairly easily computed. Knowing the mass of one mol of the compound and the mass of each element in that quantity of compound allows calculations of the percent by mass of each element.

For example, one mole of H_2O contains two mol of H and one mol of O. The formula mass is 18.0 amu; hence, there is 18.0 g/mol H_2O. Two moles of H have a mass of 2.00 g. The percent H is therefore

$$\text{Percent H} = \frac{2.00}{18.0} \times 100\% = 11.1\%\text{H}$$

In laboratory work, the identity of a compound may be established by determining its percent composition experimentally and then comparing the results with the percent composition calculated from its formula.

Empirical Formulas

The formula of a compound gives the relative number of atoms of the different elements present. It also gives the relative number of moles of the different elements present. If the formula is not known, it may be deduced from the experimentally determined composition. This procedure is possible because once the relative masses of the elements are found, the relative numbers of moles of each may be determined. Formulas derived in this manner are called **empirical formulas**. In solving a problem in which percent composition is given, any size sample may be considered, since the percentage of each element does not depend on the size of the sample.

Simplify!

The most convenient size to consider is 100 g, for with that size sample, the percentage of each element is equal to the same number of grams.

If more than two elements are present, divide all the numbers of moles by the smallest to attempt to get an integral ratio. Even after this step, it might be necessary to multiply every result by a small integer to get integral ratios, corresponding to the empirical formula.

For example, to determine the empirical formula of a sample of a compound that contains 79.59 g Fe and 30.40 g O, first calculate how many moles of Fe and O are present.

$$79.59 \text{ g Fe}\left(\frac{1 \text{ mol Fe}}{55.85 \text{ g Fe}}\right) = 1.425 \text{ mol Fe}$$

$$30.40 \text{ g O}\left(\frac{1 \text{ mol O}}{16.00 \text{ g O}}\right) = 1.900 \text{ mol O}$$

Dividing each by 1.425 (the smallest) yields 1.000 mol Fe for every 1.333 mol O. This ratio is still not integral. Multiplying these values by 3 yields integers.

3.000 mol Fe and 3.999 or 4 mol O

Thus, the empirical formula is Fe_3O_4.

Molecular Formulas

Formulas describe the composition of compounds. Empirical formulas give the mole ratio of the various elements. However, sometimes different compounds have the same ratio of moles of atoms of the same elements. For example, acetylene, C_2H_2, and benzene, C_6H_6, each have 1:1 ratios of moles of carbon atoms to moles of hydrogen atoms. Such compounds have the same percent compositions. However, they do not have the same number of atoms in each molecule. The **molecular formula** is a formula that gives the mole ratios of the various elements plus the information of how many atoms are in each molecule. In order to deduce molecular formulas from experimental data, the percent composition and the molar mass are usually determined. The molar mass may be determined in several ways, one of which will be described in Chapter 9.

It is apparent that the compounds C_2H_2 and C_6H_6 have different molecular masses. That of C_2H_2 is 26 amu; that of C_6H_6 is 78 amu. It is easy to determine the molecular mass from the molecular formula, but how can the molecular formula be determined from the empirical formula and the molecular mass? Using benzene as an example, follow these steps:

1. Determine the formula mass corresponding to one empirical formula unit. CH has a formula mass of 12 + 1 = 13 amu.
2. Divide the molecular mass by the empirical formula mass. Molecular mass = 78 amu; empirical mass = 13 amu. 78/13 = 6.
3. Multiply the number of atoms of each element in the empirical formula by the whole number found in step 2. $(CH)_6 = C_6H_6$.

Solved Problems

Solved Problem 6.1 What is the formula mass of $BaCO_3$?

Solution: The atomic mass of barium = 137.34 amu. The atomic mass of carbon = 12.01 amu. The atomic mass of oxygen is 16.00 amu, but there are three, so multiply by 3 = 48.00. The formula mass is the total (137.34 + 12.01 + 48.00) = 197.35 amu.

Solved Problem 6.2 Calculate the mass of 1.00 mol of CCl_4.

Solution: The mass contributed by C = 12.0 amu. The mass contributed by Cl_4 = 4 × 35.5 amu = 142 amu. Thus, the formula mass is the total (12.0 + 142) = 154 amu. Thus, the mass of 1.00 mol of CCl_4 is 154 g.

Solved Problem 6.3 Calculate the mass of 2.50 mol $NaClO_3$.

Solution: The mass contributed by Na = 23.0 amu. The mass contributed by Cl = 35.5 amu. The mass contributed by O_3 = 3 × 16.0 = 48.0 amu.

Thus, the formula mass of $NaClO_3$ = 23.0 + 35.5 + 48.0 = 106.5 amu. However, this problem asked for the mass of 2.50 mol, thus

$$2.50 \text{ mol NaClO}_3\left(\frac{106.5 \text{ g NaClO}_3}{1 \text{ mol NaClO}_3}\right) = 266 \text{ g NaClO}_3$$

Solved Problem 6.4. How many moles of hydrogen atoms are present in 4.00 mol of NH_3?

Solution: The formula states that there are 3 mol H for every 1 mol NH_3. Therefore,

$$4.00 \text{ mol NH}_3\left(\frac{3 \text{ mol H}}{1 \text{ mol NH}_3}\right) = 12.0 \text{ mol H}$$

Solved Problem 6.5 Calculate the percent composition of $MgSO_3$.

Solution: One mole of the compound contains 1 mol of magnesium, 1 mol of sulfur, and 3 mol of oxygen atoms. The formula mass is 104.37 amu; hence there is 104.37 g/mol $MgSO_3$.

One mole of magnesium has a mass of 24.31 g. The percent magnesium is therefore

$$\text{Percent Mg} = \frac{24.31 \text{ g Mg}}{104.37 \text{ g MgSO}_3} \times 100\% = 23.29\% \text{ Mg}$$

One mole of sulfur has a mass of 32.06 g. The percent sulfur is given by

$$\text{Percent S} = \frac{32.06 \text{ g S}}{104.37 \text{ g MgSO}_3} \times 100\% = 30.72\% \text{ S}$$

Three moles of oxygen have a mass of 3 × 16.00 g. The percent of oxygen is given by

$$\text{Percent O} = \frac{3 \times 16.00 \text{ g O}}{104.37 \text{ g MgSO}_3} \times 100\% = 45.99\% \text{ O}$$

The total of all the percentages in the compound is 100.00 percent. This result may be interpreted to mean that if there were 100.00 g of $MgSO_3$, then 23.29 g would be magnesium, 30.72 g would be sulfur, and 45.99 g would be oxygen.

Solved Problem 6.6 Determine the empirical formula of a compound containing 41.1 percent K, 33.7 percent S, and 25.2 percent O.

Solution:

$$41.1\text{ g K}\left(\frac{1\text{ mol K}}{39.1\text{ g K}}\right) = 1.05\text{ mol K}\,; \quad \frac{1.05\text{ mol K}}{1.05} = 1.00\text{ mol K}$$

$$33.7\text{ g S}\left(\frac{1\text{ mol S}}{32.06\text{ g S}}\right) = 1.05\text{ mol S}\,; \quad \frac{1.05\text{ mol S}}{1.05} = 1.00\text{ mol S}$$

$$25.2\text{ g O}\left(\frac{1\text{ mol O}}{16.0\text{ g O}}\right) = 1.58\text{ mol O}\,; \quad \frac{1.58\text{ mol O}}{1.05} = 1.5\text{ mol O}$$

Multiplying each value by 2 yields 2 mol K, 2 mol S, and 3 mol O, corresponding to $K_2S_2O_3$.

Solved Problem 6.7 A compound contains 85.7 percent carbon and 14.3 percent hydrogen and has a molar mass of 56.0 g/mol. What is its molecular formula?

Solution: The first step is to determine the empirical formula from the percent composition data.

$$85.7\text{ g C}\left(\frac{1\text{ mol C}}{12.0\text{ g C}}\right) = 7.14\text{ mol C}$$

$$14.3\text{ g H}\left(\frac{1\text{ mol H}}{1.008\text{ g H}}\right) = 14.2\text{ mol H}$$

The empirical formula is CH_2. The mass of CH_2 = 14.0 amu. The number of units = (56.0 amu)/(14.0 amu) = 4. Thus, the molecular formula = $(CH_2)_4 = C_4H_8$.

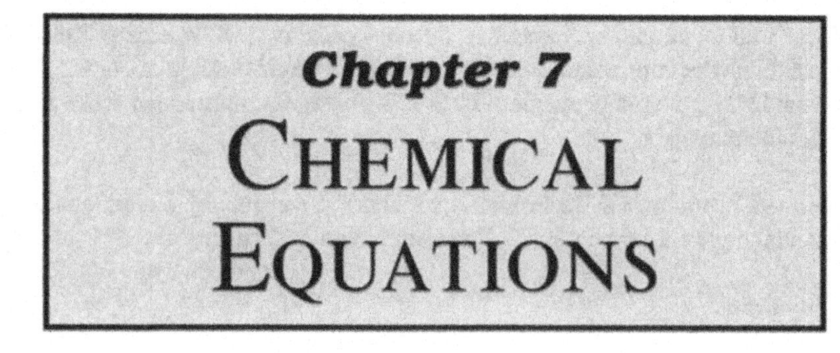

IN THIS CHAPTER:

- ✔ *Chemical Equations*
- ✔ *Balancing Simple Equations*
- ✔ *Predicting the Products of a Reaction*
- ✔ *Writing Net Ionic Equations*
- ✔ *Solved Problems*

Chemical Equations

A chemical reaction is described by means of a shorthand notation called a **chemical equation**. One or more substances, called **reactants** or **reagents**, are allowed to react to form one or more other substances, called **products**. Instead of using words, equations are written using the formulas for the substances involved. For example, a reaction used to prepare oxygen may be described in words as follows:

Mercury(I) oxide, when heated, yields oxygen gas plus mercury.

Using the formulas for the substances involved, the process could be written

$$2\,Hg_2O \xrightarrow{heat} O_2 + 4\,Hg$$

A chemical equation describes a chemical reaction in many ways as an empirical formula describes a chemical compound. The equation describes not only which substances react, but also the relative number of moles of each reactant and product. Note especially that it is the mole ratios in which the substances react, not how much is present, that the equation describes.

To show the quantitative relationships, the equation must be **balanced**. That is, it must have the same number of atoms of each element used up and produced (except for special equations that describe nuclear reactions). The law of conservation of mass is obeyed as well as the law of conservation of atoms. **Coefficients** are used before the formulas for elements and compounds to tell how many formula units of that substance are involved in the reaction. The number of atoms involved in each formula unit is multiplied by the coefficient to get the total number of atoms of each element involved. When equations with individual ions are written, the net charge on each side of the equation, as well as the numbers of atoms of each element, must be the same to have a balanced equation. The absence of a coefficient in a balanced equation implies a coefficient of one.

Balancing Simple Equations

If you know the reactants and products of a chemical reaction, you should be able to write an equation for the reaction and balance it. In writing the equation, first write the correct formulas for all reactants and products. After they are written, only then start to balance the equation. Do not balance the equation by changing the formulas or substances involved.

Before the equation is balanced, there are no coefficients for any reactant or product. To prevent ambiguity while balancing the equation, place a question mark in front of every substance. Assume a coefficient of one for the most complicated substance in the equation. Then, work from this substance to figure out the coefficient of the others, one at a time.

Replace each question mark as you figure out each coefficient. If an element appears in more than one reactant or product, leave that element for last. If a polyatomic ion is involved that does not change during the reaction, you may treat the whole thing as one unit, instead of considering the atoms that make it up. After you have provided a coefficient for all the substances, if any fractions are present multiply every coefficient by the same small integer to clear the fractions.

For example, balance the following equation:

$$CoF_3 + KI \rightarrow KF + CoI_2 + I_2$$

First, assume a coefficient of one for the most complicated reactant or product and insert question marks for the other coefficients:

$$1\ CoF_3 + ?\ KI \rightarrow ?\ KF + ?\ CoI_2 + ?\ I_2$$

Replace each question mark with a coefficient one at a time.

$$1\ CoF_3 + 3\ KI \rightarrow 3\ KF + 1\ CoI_2 + \tfrac{1}{2}\ I_2$$

Since a fraction is present, multiply every coefficient by the same small integer to clear the fractions. In this example, multiply by two. The balanced equation is:

$$2\ CoF_3 + 6\ KI \rightarrow 6\ KF + 2\ CoI_2 + I_2$$

Remember

Check to see that you have the same number of atoms of each element on both sides of the equation after you are finished.

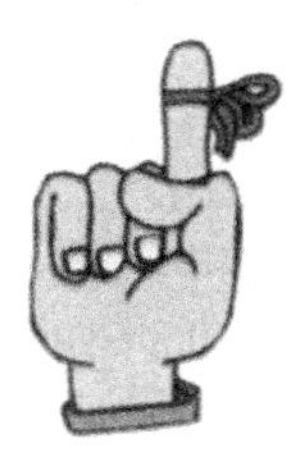

Predicting the Products of a Reaction

Before you can balance an equation, you have to know the formulas for all the reactants and products. If the names are given for these substances, you have to know how to write the formulas from the names (Chapter 5).

If only the reactants are given, you have to know how to predict the products from the reactants. Chemical reactions may be simply classified into five types:

Type I:	combination reaction
Type II:	decomposition reaction
Type III:	substitution reaction
Type IV:	double substitution reaction
Type V:	combustion reaction

Combination Reactions. A **combination reaction** is a reaction of two reactants to produce one product. The simplest combination reactions are the reactions of two elements to form a compound. For example,

$$2\,Ca + O_2 \rightarrow 2\,CaO$$

It is possible for an element and a compound of that element or for two compounds containing a common element to react by combination. The most common type in general chemistry is the reaction of a metal oxide with a nonmetal oxide to produce a salt with an oxyanion. For example,

$$CaO + SO_3 \rightarrow CaSO_4$$

Decomposition Reactions. Decomposition reactions are easy to recognize since they only have one reactant. A type of energy, such as heat or electricity, may also be indicated. The reactant decomposes to its elements, to an element and a compound, or to two simpler compounds. A **catalyst** is a substance that speeds up a chemical reaction without undergoing a permanent change in its own composition. Catalysts are often noted above or below the arrow in the chemical equation. Since a small quantity of catalyst is sufficient to cause a large quantity of reaction, the amount of catalyst need not be specified; it is not balanced as the reactants and products are. In this manner, the equation for a common laboratory preparation of oxygen is written

$$2\,KClO_3 \xrightarrow{MnO_2} 2\,KCl + 3\,O_2$$

Substitution or Replacement Reactions. Elements have varying abilities to combine. Among the most reactive metals are the alkali metals and the alkaline earth metals. Among the most stable metals are silver and gold, prized for the lack of reactivity.

When a free element reacts with a compound of different elements, the free element will replace one of the elements in the compound if the free element is more reactive than the element it replaces. In general, a free metal will replace the metal in the compound, or a free nonmetal will replace the nonmetal in the compound. A new compound and a new free element are produced. For example,

$$2\ Na + NiCl_2 \rightarrow 2\ NaCl + Ni$$

If a free element is less active than the corresponding element in the compound, no reaction will take place. A short list of metals and nonmetals in order of their reactivities is presented in Table 7.1. The metals in the list range from very active (at the top) to very stable (at the bottom); the nonmetals listed range from very active to fairly active.

Metals	Nonmetals
Alkali and alkaline earth metals	F
Al	O
Zn	Cl
Fe	Br
Pb	I
H	
Cu	
Ag	
Au	

Table 7.1 Relative reactivities of some metals and nonmetals

In substitution reactions with acids, metals that can form two different ions in their compounds generally form the one with the lower charge. For example, iron can form Fe^{2+} and Fe^{3+}. In its reaction with HCl, $FeCl_2$ is formed. In contrast, in combination with the free element, the higher-charged ion is often formed if sufficient nonmetal is available. For example,

$$2\ Fe + 3\ Cl_2 \rightarrow 2\ FeCl_3$$

Double Substitution or Double-Replacement Reactions. Double substitution or **double-replacement reactions** (also called **double-decomposition** or **metathesis reactions**) involve two ionic compounds, most often in aqueous solution. In this type of reaction, the cations simply swap anions. The reaction proceeds if a solid or a covalent compound is formed from ions in solution. All gases at room temperature are covalent. Some reactions of ionic solids plus ions in solution also occur. Otherwise no reaction takes place.

Since it is useful to know what state each reagent is in, we often designate the state in the equation. The designations are **(s)** for solid, **(l)** for liquid, **(g)** for gas, and **(aq)** for aqueous. Thus, a reaction of two ionic compounds, silver nitrate with sodium chloride in aqueous solution, yielding solid silver chloride and aqueous sodium nitrate, may be written as

$$AgNO_3(aq) + NaCl(aq) \rightarrow AgCl(s) + NaNO_3(aq)$$

Just as with replacement reactions, double-replacement reactions may or may not proceed. They need a driving force such as insolubility or covalence. In order to predict if a double replacement reaction will proceed, you must know some solubilities of ionic compounds. A short list is given in Table 7.2.

Soluble	Insoluble
Chlorates	$BaSO_4$
Acetates	Most sulfides
Nitrates	Most oxides
Alkali metal salts	Most carbonates
Ammonium salts	Most phosphates
Chlorides, except those listed...	$AgCl$, $PbCl_2$, Hg_2Cl_2, $CuCl$

Table 7.2 Some solubility classes

In double-replacement reactions, the charges on the metal ions (and nonmetal ions if they do not form covalent compounds) generally remain throughout the reaction.

NH_4OH and H_2CO_3 are unstable. If one of these products were expected as a product, either NH_3 plus H_2O or CO_2 plus H_2O would be obtained instead.

Combustion Reactions. Reactions of elements and compounds with oxygen are so prevalent that they may be considered a separate type of reaction, a **combustion reaction**. Compounds of carbon, hydrogen, oxygen, sulfur, nitrogen, and other elements may be burned. If a reactant contains carbon, then carbon monoxide or carbon dioxide will be produced, depending on how much oxygen is available. Reactants containing hydrogen always produce water on burning. NO and SO_2 are other products of burning. A catalyst is required to produce SO_3 in a combustion reaction with O_2.

Acids and Bases. Generally, acids react according to the rules for replacement and double replacement reactions. They are so important, however, that a special nomenclature has developed for acids and their reactions. Acids were introduced in Chapter 5. They may be identified by their formulas, which have the H representing hydrogen written first, and by their names, which contain the word "acid." An acid will react with a base to form a salt and water. The process is called **neutralization**. The driving force for such reactions is the formation of water, a covalent compound. For example,

$$HBr + NaOH \rightarrow NaBr + H_2O$$

Writing Net Ionic Equations

When a substance made up of ions is dissolved in water, the dissolved ions undergo their own characteristic reactions regardless of what other ions may be present. For example, barium ions in solution always react with sulfate ions in solution to form an insoluble ionic compound,

$BaSO_4$(s), no matter what other ions are present in the barium solution. If solutions of barium chloride and sodium sulfate are mixed, a white solid, barium sulfate, is produced. The solid can be separated from the solution by filtration, and the resulting solution contains sodium chloride, just as it would if solid NaCl were added to water.

$$BaCl_2 + Na_2SO_4 \rightarrow BaSO_4(s) + 2\ NaCl$$

or

$$Ba^{2+} + 2\ Cl^- + 2\ Na^+ + SO_4^{2-} \rightarrow BaSO_4(s) + 2\ Na^+ + 2\ Cl^-$$

The latter equation shows that in effect, the sodium ions and the chloride ions have not changed. They began as ions in solution and wound up as those same ions in solution. They are called **spectator ions**. Since they have not reacted, it is not really necessary to include them in the equation. If they are left out, a **net ionic equation** results:

$$Ba^{2+} + SO_4^{2-} \rightarrow BaSO_4(s)$$

Net ionic equations may be written whenever reactions occur in solution in which some of the ions originally present are removed from solution or when ions not originally present are formed. Usually, ions are removed from solution by one of the following processes:

1. Formation of an insoluble ionic compound (see Table 7.2).
2. Formation of molecules containing only covalent bonds.
3. Formation of new ionic species.
4. Formation of a gas.

The following generalizations may help in deciding whether a compound is ionic or covalent.

1. Binary compounds of two nonmetals are covalently bonded. However, strong acids in water form ions completely.

2. Binary compounds of a metal and nonmetal are usually ionic.

3. Ternary compounds are usually ionic, at least in part, except if they contain no metal atoms or ammonium ion.

Net ionic equations must always have the same net charge on each side of the equation. The same number of each type of spectator ion must be omitted from both sides of the equation.

Solved Problems

Solved Problem 7.1 Zinc metal reacts with HCl to produce $ZnCl_2$ and hydrogen gas. Write a balanced equation for the process.

Solution: Start by writing the correct formulas for all reactants and products. Put question marks in the place of all the coefficients except the most complicated substance ($ZnCl_2$); put a 1 in front of that substance.

$$? \text{ Zn} + ? \text{ HCl} \rightarrow 1 \text{ ZnCl}_2 + ? \text{ H}_2$$

Note that hydrogen is one of the seven elements that form diatomic molecules when in the elemental state. Work from $ZnCl_2$, and balance the other elements one at a time.

$$1 \text{ Zn} + ? \text{ HCl} \rightarrow 1 \text{ ZnCl}_2 + ? \text{ H}_2$$
$$1 \text{ Zn} + 2 \text{ HCl} \rightarrow 1 \text{ ZnCl}_2 + ? \text{ H}_2$$
$$1 \text{ Zn} + 2 \text{ HCl} \rightarrow 1 \text{ ZnCl}_2 + 1 \text{ H}_2$$

Since the coefficient one is implied if there are no coefficients, remove the ones to simplify.

$$\text{Zn} + 2 \text{ HCl} \rightarrow \text{ZnCl}_2 + \text{H}_2$$

There are one Zn atom, two H atoms, and two Cl atoms on each side.

Solved Problem 7.2 Write a complete, balanced equation for the reaction that occurs when $MgCO_3$ is heated.

Solution: This is a decomposition reaction. A ternary compound decomposes into two simpler components. Note that energy is added to cause this reaction

$$MgCO_3 \xrightarrow{\text{heat}} MgO + CO_2$$

Solved Problem 7.3 Complete and balance the following equation. If no reaction occurs, indicate that fact.

$$Al + HCl \rightarrow$$

Solution: Aluminum is more reactive than hydrogen (see Table 7.1) and replaces it from its compounds. Note that free hydrogen is in the form H_2.

$$2\ Al + 6\ HCl \rightarrow 2\ AlCl_3 + 3\ H_2$$

Solved Problem 7.4 Complete and balance the following equation. If no reaction occurs, indicate that fact.

$$FeCl_2 + AgNO_3 \rightarrow$$

Solution: This is a double displacement reaction. If you start with Fe^{2+}, you wind up with Fe^{2+}.

$$FeCl_2(aq) + 2\ AgNO_3(aq) \rightarrow Fe(NO_3)_2(aq) + 2\ AgCl(s)$$

Solved Problem 7.5 Complete and balance the following equation.

$$C_2H_4 + O_2 \text{ (limited amount)} \rightarrow$$

Solution: This is a combustion reaction. CO_2 is produced only when there is sufficient O_2 available (3 mol O_2 per mole C_2H_4).

$$C_2H_4 + 2\ O_2 \rightarrow 2\ CO + 2\ H_2O$$

Solved Problem 7.6 What type of chemical reaction is represented by each of the following? Complete and balance the equation for each.

(*a*) $Cl_2 + NaBr \rightarrow$
(*b*) $Cl_2 + K \rightarrow$

(*c*) $CaCO_3 \xrightarrow{\text{heat}}$

(*d*) $ZnCl_2 + AgC_2H_3O_2 \rightarrow$

(*e*) $C_3H_8 + O_2$ (excess) $\rightarrow$

Solution:

(*a*) substitution	$Cl_2 + 2\ NaBr \rightarrow 2\ NaCl + Br_2$
(*b*) combination	$Cl_2 + 2\ K \rightarrow 2\ KCl$
(*c*) decomposition	$CaCO_3 \xrightarrow{\text{heat}} CO_2 + CaO$
(*d*) double substitution	$ZnCl_2 + 2\ AgC_2H_3O_2 \rightarrow Zn(C_2H_3O_2)_2 + 2\ AgCl$
(*e*) combustion	$C_3H_8 + 5\ O_2 \rightarrow 3\ CO_2 + 4\ H_2O$

Solved Problem 7.7 Predict which of the following will contain ionic bonds: (*a*) $CoCl_2$, (*b*) CO, (*c*) CaO, (*d*) NH_4Cl, (*e*) H_2SO_4, (*f*) HCl, and (*g*) SCl_2.

Solution: (*a*) $CoCl_2$, (*c*) CaO, and (*d*) NH_4Cl contain ionic bonds. NH_4Cl also has covalent bonds within the ammonium ion. (*e*) H_2SO_4 and (*f*) HCl would form ions if allowed to react with water.

Solved Problem 7.8 Write a net ionic equation for the reaction of aqueous $Ba(OH)_2$ with aqueous HCl.

Solution: The overall equation is

$$Ba(OH)_2 + 2\ HCl \rightarrow BaCl_2 + 2\ H_2O$$

In ionic form:

$$Ba^{2+} + 2\ OH^- + 2\ H^+ + 2\ Cl^- \rightarrow Ba^{2+} + 2\ Cl^- + 2\ H_2O$$

Leaving out the spectator ions and dividing each side by two yields

$$OH^- + H^+ \rightarrow H_2O$$

Solved Problem 7.9 Write a net ionic equation for each of the following overall reactions:

(*a*) $H_3PO_4 + 2\,NaOH \rightarrow Na_2HPO_4 + 2\,H_2O$
(*b*) $CaCO_3(s) + CO_2 + H_2O \rightarrow Ca(HCO_3)_2$
(*c*) $NaHCO_3 + NaOH \rightarrow Na_2CO_3 + H_2O$

Solution:

(*a*) $H_3PO_4 + 2\,OH^- \rightarrow HPO_4^{2-} + 2\,H_2O$
(*b*) $CaCO_3 + CO_2 + H_2O \rightarrow Ca^{2+} + 2\,HCO_3^-$
(*c*) $HCO_3^- + OH^- \rightarrow CO_3^{2-} + H_2O$

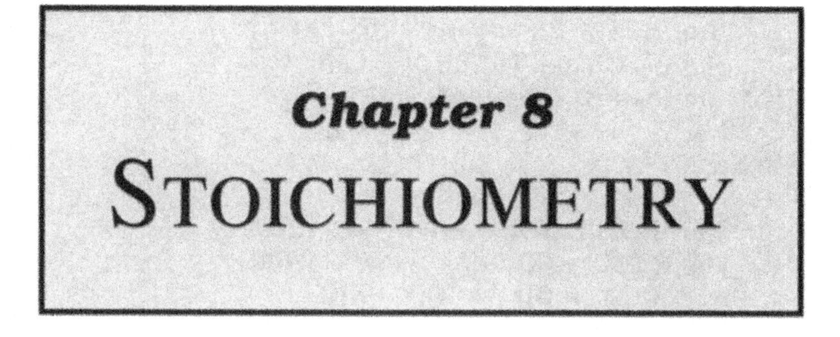

Chapter 8 STOICHIOMETRY

IN THIS CHAPTER:

- ✔ *Mole-to-Mole Calculations*
- ✔ *Limiting Quantities*
- ✔ *Calculations Based on Net Ionic Equations*
- ✔ *Heat Capacity and Heat of Reaction*
- ✔ *Solved Problems*

Mole-to-Mole Calculations

In chemical work, it is important to be able to calculate how much raw material is needed to prepare a certain quantity of products. It is also useful to know if a certain reaction method can prepare more product from a given quantity of material than another reaction method. Analyzing material means finding out how much of each element is present. To do the measurements, parts of the material are often converted to compounds that are easy to separate, and then those compounds are measured. All these measurements involve **stoichiometry**, the science of measuring how much of one thing can be produced from certain amounts of others. Calculations involving stoichiometry are also used in studying the gas laws, solution chemistry, equilibrium, and other topics.

Balanced chemical equations express ratios of numbers of formula units of each chemical involved in a reaction. They may also be used to express the ratio of moles of reactants and products. For the reaction

$$N_2 + 3\ H_2 \rightarrow 2\ NH_3$$

one mole of N_2 reacts with three moles of H_2 to produce two moles of NH_3. To determine how many moles of nitrogen it takes to react with 1.97 mol of hydrogen, simply set up an equation using the ratios:

$$1.97\ \text{mol H}_2 \left(\frac{1\ \text{mol N}_2}{3\ \text{mol H}_2} \right) = 0.657\ \text{mol N}_2$$

The balanced equation expresses quantities in moles, but it is seldom possible to measure out quantities in moles directly. If the quantities given or required are expressed in other units, it is necessary to convert them to moles before using the factors of the balanced chemical equation. Conversion of mass to moles and vice versa was considered in Chapter 6. Not only mass, but any measurable quantity that can be converted to moles may be treated in this manner to determine the quantity of product or reactant involved in a reaction from the quantity of any other reactant or product.

Limiting Quantities

If nothing is stated about the quantity of a reactant in a reaction, it must be assumed to be present in sufficient quantity to allow the reaction to take place. Sometimes the quantity of only one reactant is given, and you may assume that the other reactants are present in sufficient quantity. However, other times, the quantities of more than one reactant will be stated. This type of problem is called a **limiting-quantities problem**.

To solve a limiting quantities problem in which the reactant in excess is not obvious, do as follows:

1. Calculate the number of moles of one reactant required to react with all the other reactant present.
2. Compare the number of moles of the one reactant that is present and the number of moles required. This comparison will tell you which reactant is present in excess and which one is in limiting quantity.

3. Calculate the quantity of reaction (reactants used up and products produced) on the basis of the quantity of reactant in limiting quantity.

For example, to determine how many moles of NaCl can be produced by the reaction of 2.0 mol NaOH and 3.0 mol HCl, first write a balanced equation.

$$\text{NaOH} + \text{HCl} \rightarrow \text{NaCl} + \text{H}_2\text{O}$$

Next, determine the number of moles of NaOH required to react completely with 3.0 mol of HCl:

$$3.0 \text{ mol HCl}\left(\frac{1 \text{ mol NaOH}}{1 \text{ mol HCl}}\right) = 3.0 \text{ mol NaOH required}$$

Since there is 2.0 mol NaOH present, but the HCl present would require 3.0 mol NaOH, there is not enough NaOH to react completely with the HCl. The NaOH is present in limiting quantity. Now, the number of moles of NaCl that can be produced is calculated on the basis of the 2.0 mol NaOH present:

$$2.0 \text{ mol NaOH}\left(\frac{1 \text{ mol NaCl}}{1 \text{ mol NaOH)}}\right) = 2.0 \text{ mol NaCl}$$

Alternatively, the problem could have been started by calculating the quantity of HCl required to react completely with the NaOH present:

$$2.0 \text{ mol NaOH}\left(\frac{1 \text{ mol HCl}}{1 \text{ mol NaOH}}\right) = 2.0 \text{ mol HCl required}$$

Since 2.0 mol HCl is required to react with all the NaOH and there is 3.0 mol of HCl present, HCl present in excess. If HCl is in excess, NaOH must be limiting. It is not necessary to do both calculations. The same result will be obtained no matter which is used. If the quantities of both reactants are in exactly the correct ratio for the balanced chemical equation, then either reactant maybe used to calculate the quantity of product produced.

Remember

If you are given the molar concentration of a solution, the number of moles is simply volume × molar concentration.

Calculations Based on Net Ionic Equations

The net ionic equation, like all balanced chemical equations, gives the ratio of moles of each substance to moles of each of the others. It does not immediately yield information about the mass of the entire salt, however. (One cannot weigh out only Ba^{2+} ions.) Therefore, when masses of reactants are required, the specific compound used must be included in the calculation.

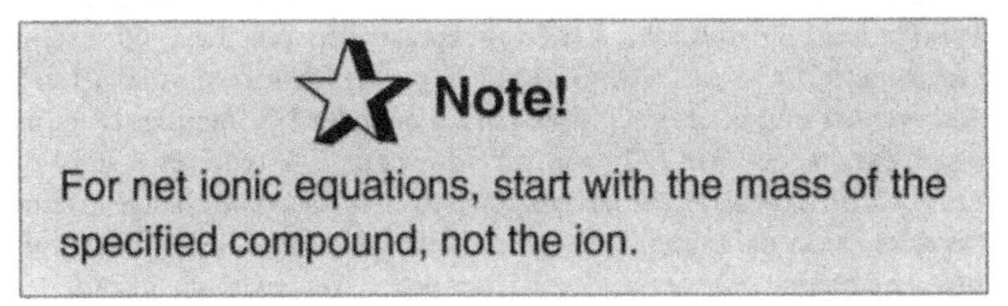

Note!

For net ionic equations, start with the mass of the specified compound, not the ion.

For example, to determine the mass of silver nitrate required to prepare 100 g AgCl, first determine how many moles of silver ion are required to make 100 g AgCl. Write the net ionic equation:

$$Ag^+ + Cl^- \rightarrow AgCl$$

The molar mass of AgCl is 108 + 35 = 143 g/mol. In 100 g of AgCl, which is to be prepared, there is

$$100 \text{ g AgCl}\left(\frac{1 \text{ mol AgCl}}{143 \text{ g AgCl}}\right) = 0.699 \text{ mol AgCl}$$

Hence, from the balanced chemical equation, 0.699 mol of Ag^+ is required:

$$0.699 \text{ mol AgCl}\left(\frac{1 \text{ mol Ag}^+}{1 \text{ mol AgCl}}\right) = 0.699 \text{ mol Ag}^+$$

The 0.699 mol of Ag^+ may be furnished from 0.699 mol of $AgNO_3$. Then

$$0.699 \text{ mol AgNO}_3\left(\frac{170 \text{ g AgNO}_3}{1 \text{ mol AgNO}_3}\right) = 119 \text{ g AgNO}_3$$

Heat Capacity and Heat of Reaction

Heat is a reactant or product in most chemical reactions. Before we consider including heat in a balanced chemical equation, first we must learn how to measure heat. When heat is added to a system in the absence of a chemical reaction the system may warm up or a change of phase may occur. In this section, only the warming process will be considered.

Temperature is a measure of the intensity of the energy in a system. The **specific heat capacity** of a substance is the quantity of heat required to heat 1 g of the substance 1 °C. Specific heat capacity is often called **specific heat**. Lowercase ***c*** is used to represent specific heat. For example, the specific heat of water is 4.184 J/(g · °C). This means that 4.184 J will warm 1 g of water 1 °C. To warm 2 g of water 1 °C requires twice as much energy, or 8.368 J. To warm 1 g of water 2 °C requires 8.368 J of energy also. In general, the heat required to effect a certain change in temperature in a certain sample of a given material is calculated with the following equation, where the Greek letter delta (Δ) means "change in."

$$\text{Heat required} = (\text{mass})(\text{specific heat})(\text{change in temperature}) = (m)(c)(\Delta t)$$

Know the Difference!

Temperature ≠ Heat

Solved Problems

Solved Problem 8.1 Sulfuric acid reacts with sodium hydroxide to produce sodium sulfate and water. (*a*) Write a balanced chemical equation for the reaction. (*b*) Determine the number of moles of sulfuric acid in 50.0 g sulfuric acid. (*c*) How many moles of sodium sulfate will be produced by the reaction of this number of moles of sulfuric acid? (*d*) How many grams of sodium sulfate will be produced? (*e*) How many moles of sodium hydroxide will it take to react with this quantity of sulfuric acid? (*f*) How many grams of sodium hydroxide will be used up?

Solution:

(*a*) $H_2SO_4 + 2\,NaOH \rightarrow Na_2SO_4 + 2\,H_2O$

(*b*) $50.0\text{ g }H_2SO_4\left(\dfrac{1\text{ mol }H_2SO_4}{98.0\text{ g }H_2SO_4}\right) = 0.510\text{ mol }H_2SO_4$

(*c*) $0.510\text{ mol }H_2SO_4\left(\dfrac{1\text{ mol }Na_2SO_4}{1\text{ mol }H_2SO_4}\right) = 0.510\text{ mol }Na_2SO_4$

(*d*) $0.510\text{ mol }Na_2SO_4\left(\dfrac{142\text{ g }Na_2SO_4}{1\text{ mol }Na_2SO_4}\right) = 72.4\text{ g }Na_2SO_4$

(*e*) $0.510\text{ mol }H_2SO_4\left(\dfrac{2\text{ mol NaOH}}{1\text{ mol }Na_2SO_4}\right) = 1.02\text{ mol NaOH}$

(*f*) $1.02\text{ mol NaOH}\left(\dfrac{40.0\text{ g NaOH}}{1\text{ mol NaOH}}\right) = 40.8\text{ g NaOH}$

Solved Problem 8.2 How many moles of PbI_2 can be prepared by the reaction of 0.252 mol of $Pb(NO_3)_2$ and 0.452 mol NaI?

Solution: The balanced equation is

$$Pb(NO_3)_2 + 2\,NaI \rightarrow PbI_2 + 2\,NaNO_3$$

First, determine how many moles of NaI are required to react with all the $Pb(NO_3)_2$ present:

$$0.252 \text{ mol Pb(NO}_3)_2 \left(\frac{2 \text{ mol NaI}}{1 \text{ mol Pb(NO}_3)_2}\right) = 0.504 \text{ mol NaI required}$$

Since more NaI is required (0.504 mol) than is present (0.452 mol), NaI is in limiting quantity.

$$0.452 \text{ mol NaI} \left(\frac{1 \text{ mol PbI}_2}{2 \text{ mol NaI}}\right) = 0.226 \text{ mol PbI}_2$$

Note especially that the number of moles of NaI exceeds the number of moles of $Pb(NO_3)_2$ present, but that these numbers are not what must be compared. Compare the number of moles of one reactant present with the number of moles of that same reactant required! The ratio of moles of NaI to $Pb(NO_3)_2$ in the equation is 2 : 1, but in the reaction mixture that ratio is less than 2 : 1; therefore, the NaI is in limiting quantity.

Solved Problem 8.3 How many grams of $Ca(ClO_4)_2$ can be prepared by treatment of 22.5 g CaO with 125 g $HClO_4$?

Solution: The balanced equation is

$$\text{CaO} + 2\ \text{HClO}_4 \rightarrow \text{Ca(ClO}_4)_2 + \text{H}_2\text{O}$$

This problem gives quantities of the two reactants in grams; we must first change them to moles:

$$22.5 \text{ g CaO} \left(\frac{1 \text{ mol CaO}}{56.0 \text{ g CaO}}\right) = 0.402 \text{ mol CaO}$$

$$125.0 \text{ g HClO}_4 \left(\frac{1 \text{ mol HClO}_4}{100 \text{ g HClO}_4}\right) = 1.25 \text{ mol HClO}_4$$

Next, determine the limiting reactant:

$$1.25 \text{ mol HClO}_4 \left(\frac{1 \text{ mol CaO}}{2 \text{ mol HClO}_4}\right) = 0.625 \text{ mol CaO required}$$

Since 0.625 mol CaO is required and 0.402 mol CaO is present, CaO is in limiting quantity.

$$0.402 \text{ mol CaO}\left(\frac{1 \text{ mol Ca(ClO}_4)_2}{1 \text{ mol CaO}}\right)\left(\frac{239 \text{ g Ca(ClO}_4)_2}{1 \text{ mol Ca(ClO}_4)_2}\right)$$

$$= 96.1 \text{ g Ca(ClO}_4)_2 \text{ produced.}$$

Solved Problem 8.4 What is the maximum mass of $BaSO_4$ that can be produced when a solution containing 10.0 g of Na_2SO_4 is added to another solution containing an excess of Ba^{2+}?

Solution:

$$Ba^{2+} + SO_4^{2-} \rightarrow BaSO_4$$

$$10.0 \text{ g Na}_2\text{SO}_4\left(\frac{1 \text{ mol Na}_2\text{SO}_4}{142 \text{ g Na}_2\text{SO}_4}\right)\left(\frac{1 \text{ mol SO}_4^{2-}}{1 \text{ mol Na}_2\text{SO}_4}\right)$$

$$\left(\frac{1 \text{ mol BaSO}_4}{1 \text{ mol SO}_4^{2-}}\right)\left(\frac{233 \text{ g BaSO}_4}{1 \text{ mol BaSO}_4}\right) = 16.4 \text{ g BaSO}_4$$

Solved Problem 8.5 How much heat does it take to raise the temperature of 100.0 g of water 17.0 °C?

Solution:

$$\text{Heat} = (m)(c)(\Delta t) = (100.0 \text{ g})\left(\frac{4.184 \text{ J}}{\text{g} \cdot {}^\circ\text{C}}\right)(17.0\ {}^\circ\text{C}) = 7110 \text{ J} = 7.11 \text{ kJ}$$

Solved Problem 8.6 How much heat will be produced by burning 20.0 g of carbon to carbon dioxide?

$$C + O_2 \rightarrow CO_2 + 393 \text{ kJ}$$

Solution:

$$20.0 \text{ g C}\left(\frac{1 \text{ mol C}}{12.0 \text{ g C}}\right)\left(\frac{393 \text{ kJ}}{1 \text{ mol C}}\right) = 655 \text{ kJ}$$

Chapter 9
GASES

IN THIS CHAPTER:

- ✔ *Gases*
- ✔ *Pressure of Gases*
- ✔ *Gas Laws*
- ✔ *The Combined Gas Law*
- ✔ *The Ideal Gas Law*
- ✔ *Dalton's Law of Partial Pressures*
- ✔ *Kinetic Molecular Theory*
- ✔ *Graham's Law*
- ✔ *Solved Problems*

Gases

Solid objects have definite volume and a fixed shape; **liquids** have no fixed shape other than that of their containers but do have definite volume. **Gases** have neither fixed shape nor fixed volume. Gases expand when they are heated in a nonrigid container and contract when they are cooled or subjected to increased pressure. They readily diffuse with other gases. Any quantity of gas will occupy the entire volume of its container, regardless of the size of the container.

Pressure of Gases

Pressure is defined as force per unit area. **Fluids** (liquids and gases) exert pressure in all directions. The pressure of a gas is equal to the pressure on the gas. A way of measuring pressure is by means of a **barometer**. The **standard atmosphere** (atm) is defined as the pressure that will support a column of mercury to a vertical height of 760 mm at a temperature of 0 °C. It is convenient to express the measured gas pressure in terms of the vertical height of a mercury column that the gas is capable of supporting. Thus, if the gas supports a column of mercury to a height of only 76 mm, the gas is exerting a pressure of 0.10 atm:

$$76\,\text{mm}\left(\frac{1\ \text{atm}}{760\ \text{mm}}\right) = 0.10\ \text{atm}$$

Note that the dimension 1 atm is not the same as **atmospheric pressure**. The atmospheric pressure, the pressure of the atmosphere, varies widely from day to day and from place to place, whereas the dimension 1 atm has a fixed value by definition. The unit **torr** is currently used to indicate the pressure necessary to support mercury to a vertical height of 1 mm. Thus, 1 atm = 760 torr.

Gas Laws

Robert Boyle (1627–1691) studied the effect of changing the pressure of a gas on its volume at constant temperature. He concluded that at constant temperature, the volume of a given sample of gas is inversely proportional to its pressure. This is known as **Boyle's law**. It means that as the pressure increases, the volume becomes smaller by the same factor. That is, if the pressure is doubled, the volume is halved. This relationship can be expressed mathematically by any of the following:

$$P \propto \frac{1}{V} \qquad P = \frac{k}{V} \qquad PV = k$$

Where P represents the pressure, V represents the volume, and k is a constant.

If for a given sample of gas at a given temperature, the product PV is a constant, then changing the pressure from some initial value P_1 to a new value P_2 will cause a corresponding change in the volume from the original volume V_1 to a new volume V_2 such that

$$P_1V_1 = k = P_2V_2$$
$$P_1V_1 = P_2V_2$$

The units of the constant k are determined by the units used to express the volume and the pressure.

If a given quantity of gas is heated at constant pressure in a container that has a movable wall, the volume of the gas will increase. If a given quantity of gas is heated in a container that has a fixed volume, its pressure will increase. Conversely, cooling a gas at constant pressure causes a decrease in its volume, while cooling it at constant volume causes a decrease in its pressure.

J. A. Charles (1746–1823) observed, and J. L. Gay-Lussac (1778–1850) confirmed, that when a given mass of gas is cooled at constant pressure, it shrinks by 1/273 times its volume at 0 °C for every degree Celsius that it is cooled. Conversely, when the mass of gas is heated at constant pressure, it expands by 1/273 times its volume at 0 °C for every degree Celsius that it is heated.

The chemical identity of the gas has no influence on the volume changes as long as the gas does not liquefy in the range of temperatures studied. The volume of the gas changes linearly with temperature. If it were assumed that the gas does not liquefy at very low temperatures, each sample would have zero volume at $^-273$ °C. Of course, any real gas could never have zero volume. Gases liquefy before this very cold temperature is reached. Nevertheless, $^-273$ °C is the temperature at which a sample of gas would theoretically have zero volume. Therefore, the temperature

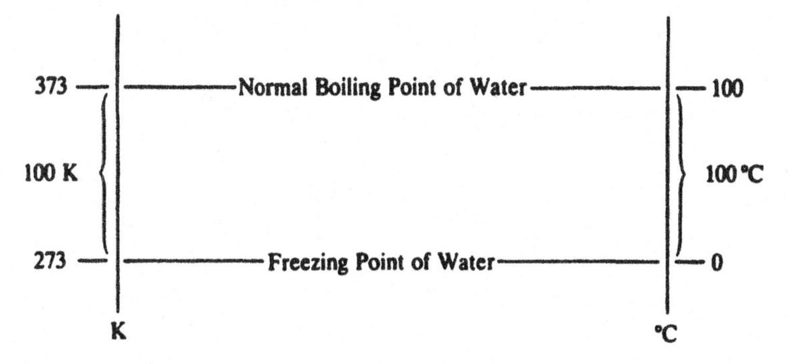

Figure 9-1 Comparison of Kelvin and Celsius temperature scales

$^-273$ °C can be regarded as the **absolute zero** of temperature. Since there cannot be less than zero volume, there can be no temperature colder than $^-273$ °C. The temperature scale that has been devised using this fact is called the **Kelvin**, or **absolute**, temperature scale. A comparison of the Kelvin scale and the Celsius scale is shown in Figure 9-1. It is seen that any temperature in degrees Celsius may be converted to Kelvins by adding 273°. It is customary to use capital T to represent Kelvin temperatures and small t to represent Celsius temperatures.

$$T = t + 273°$$

The fact that the volume of a gas varies linearly with temperature is combined with the concept of absolute temperature to give **Charles' law**: at constant pressure, the volume of a given sample of gas is directly proportional to its absolute temperature.

Expressed mathematically,

$$V = kT \quad \text{or} \quad \frac{V}{T} = k$$

Since V/T is a constant, this ratio for a given sample of gas at one volume and temperature is equal to the same ratio at any other volumes and temperatures. That is, for a given sample at constant pressure,

$$\frac{V_1}{T_1} = \frac{V_2}{T_2}$$

The Combined Gas Law

The fact that the volume V of a given mass of gas is inversely proportional to its pressure P and directly proportional to its absolute temperature T can be combined mathematically to give a single equation:

$$V = k\left(\frac{T}{P}\right) \quad \text{or} \quad \frac{PV}{T} = k$$

where k is the proportionality constant. That is, for a given mass of gas, the ratio PV/T remains constant, and therefore

$$\frac{P_1V_1}{T_1} = \frac{P_2V_2}{T_2}$$

This is the **combined gas law**. Note that if temperature is constant, the expression reduces to that for Boyle's law. If the pressure is constant, the expression is equivalent to Charles' law.

To compare quantities of gas present in two different samples, it is useful to adopt a set of standard conditions of temperature and pressure. The standard temperature is chosen as 273 K (0 °C), and the standard pressure is chosen as exactly 1 atm (760 torr). Together, these conditions are referred to as **standard conditions** or as **standard temperature and pressure (STP)**.

The Ideal Gas Law

All the gas laws described so far worked only for a given sample of gas. If a gas is produced during a chemical reaction or some of the gas under study escapes during processing, these gas laws do not apply. The **ideal gas law** works for any sample of gas. Consider a given sample of gas, for which

$$\frac{PV}{T} = k \quad \text{(a constant)}$$

If we increase the number of moles of gas at constant pressure and temperature, the volume must also increase. Thus, we can conclude that the constant k can be regarded as a product of two constants, one of which represents the number of moles of gas. We then get

$$\frac{PV}{T} = nR \quad \text{or} \quad PV = nRT$$

where n is the number of moles of gas molecules and **R** is a new constant that is valid for any sample of gas. $R = 0.0821\ \text{L} \cdot \text{atm/(mol} \cdot \text{K)}$. This equation is known as the **ideal gas law**.

In ideal gas law problems, the temperature must be given as absolute temperature, in kelvins. The units of P and V are most conveniently given in atmospheres and liters, respectively, because the units of R with the value given above are in term so these units. If other units are given, be sure to convert them.

Dalton's Law of Partial Pressures

When two or more gases are mixed, they each occupy the entire volume of the container. They each have the same temperature. However, each gas exerts its own pressure, independent of the other gases. According to **Dalton's law of partial pressures**, their pressures must add up to the total pressure of the gas mixtures. The ideal gas law applies to each individual gas in the mixture as well as to the gas mixture as a whole. In the equation

$$PV = nRT$$

Variables V, T, and R refer to each gas and the total mixture. To determine the number of moles of one gas in the mixture, use its pressure. To get the total number of moles, use the total pressure.

At 25 °C, water is ordinarily a liquid. However, in a closed container even at 25 °C, water evaporates to get a 24 torr water vapor pressure in its container. The pressure of the gaseous water is called its **vapor pressure** at that temperature. At different temperatures, it evaporates to different extents to give different vapor pressures. As long as there is liquid water present, however, the vapor pressure above pure water depends on the temperature alone. Only the nature of the liquid and the temperature affect the vapor pressure; the volume of the container does not affect its final pressure. The water vapor mixes with any other gas(es) present, and the mixture is governed by Dalton's law of partial pressure, just as any other gas mixture is.

Kinetic Molecular Theory

Under ordinary conditions of temperature and pressure, all gases are made of molecules (including one-atom molecules such as are present in the samples of the noble gases). Ionic substances do not form gases under conditions prevalent on earth. The molecules of a gas act according to the following postulates of the **kinetic molecular theory**:

1. Molecules are in constant random motion. They move in any direction until they collide with another molecule or a wall.
2. The molecules exhibit negligible intermolecular attractions or repulsions except when they collide. They move in a straight line between collisions.
3. Molecular collisions are elastic, which means that although the molecules transfer energy from one to another, as a whole they do not lose

kinetic energy when they collide. There is no friction in molecular collisions.

4. The molecules occupy a negligible fraction of the volume occupied by the gas as a whole.

5. The **average kinetic energy** of the gas molecules is directly proportional to the absolute temperature of the gas.

$$\overline{\text{KE}} = \frac{3}{2} kT = \frac{1}{2} \overline{mv^2}$$

The overbar means "average." The k in the proportionality constant is called the **Boltzmann constant**. It is equal to R, the ideal gas law constant, divided by Avogadro's number. Note that this k is the same for all gases. If two gases are at the same temperature, their molecules will have the same average kinetic energies.

Kinetic molecular theory explains why gases exert pressure. The constant bombardment of the walls of the vessel by the gas molecules causes a constant force to be applied to the wall. The force applied, divided by the area of the wall, is the pressure of the gas.

Graham's Law

Graham's law states that the rate of effusion or diffusion of a gas is inversely proportional to the square root of its molar mass. **Effusion** is the passage of a gas through small holes in its container, such as helium atoms escaping through the tiny pores of a deflating balloon over several days. **Diffusion** is the passage of a gas through another gas. For example, if a bottle of ammonia is spilled in one corner of a room, the odor of ammonia is soon apparent throughout the room. The heavier a molecule of gas, the more slowly it effuses or diffuses.

$$\frac{r_1}{r_2} = \sqrt{\frac{\text{MM}_2}{\text{MM}_1}}$$

Since two gases are at the same temperature, their average kinetic energies are the same:

$$\overline{\text{KE}_1} = \overline{\text{KE}_2} = \frac{1}{2} m_1 \overline{v_1^2} = \frac{1}{2} m_2 \overline{v_2^2}$$

Multiplying the last of these equations by 2 yields

$$m_1 \overline{v_1^2} = m_2 \overline{v_2^2} \quad \text{or} \quad \frac{m_1}{m_2} = \frac{\overline{v_2^2}}{\overline{v_1^2}}$$

Since the masses of the molecules are proportional to their molar masses, and the average velocity of the molecules is a measure of the rate of effusion or diffusion, all we have to do to this equation to get Graham's law is to take the square root. (The square root of $\overline{v^2}$ is not quite equal to the average velocity, but is a quantity called the **root mean square velocity**.)

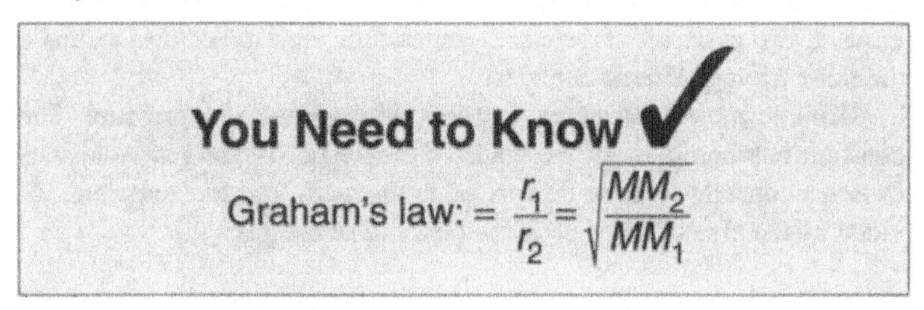

Solved Problems

Solved Problem 9.1 What is the pressure in atmospheres of a gas that supports a column of mercury to a height of 923 mm?

Solution: Pressure = 923 mm $\left(\frac{1 \text{ atm}}{760 \text{ mm}}\right) = 1.21$ atm

Solved Problem 9.2 To what pressure must a sample of gas be subjected at constant temperature in order to compress it from 500 mL to 350 mL if its original pressure is 2.11 atm?

Solution: $V_1 = 500$ mL, $P_1 = 2.11$ atm, $V_2 = 350$ mL, $P_2 = ?$

$$P_1V_1 = P_2V_2$$

$$P_2 = \frac{P_1V_1}{V_2} = \frac{(2.11 \text{ atm})(500 \text{ mL})}{350 \text{ mL}} = 3.01 \text{ atm}$$

Solved Problem 9.3 A 6.50 L sample of gas is warmed at constant pressure from 275 K to 350 K. What is its final volume?

Solution: $V_1 = 6.50$ L, $T_1 = 275$ K, $T_2 = 350$ K, $V_2 = ?$

$$\frac{V_1}{T_1} = \frac{V_2}{T_2}$$

$$V_2 = \frac{V_1 T_2}{T_1} = \frac{(6.50\text{ L})(350\text{ K})}{(275\text{ K})} = 8.27\text{ L}$$

Solved Problem 9.4 A sample of gas occupies a volume of 13.5 L at 22 °C and 1.25 atm pressure. What is the volume of this sample at STP?

Solution: $P_1 = 1.25$ atm, $V_1 = 13.5$ L, $T_1 = 22$ C + 273° = 295 K, $P_2 =$ 1.00 atm, $T_2 = 273$ K, $V_2 = ?$

$$\frac{P_1 V_1}{T_1} = \frac{P_2 V_2}{T_2}$$

$$V_2 = \frac{P_1 V_1}{T_1} \cdot \frac{T_2}{P_2} = \frac{(1.25\text{ atm})(13.5\text{ L})(273\text{ K})}{(295\text{ K})(1.00\text{ atm})} = 15.6\text{ L}$$

Solved Problem 9.5 How many moles of O_2 are present in a 1.71 L sample at 35 °C and 1.20 atm?

Solution:

$$T = 35\text{ °C} + 273 = 308\text{ K}$$

$$n = \frac{PV}{RT} = \frac{(1.20\text{ atm})(1.71\text{ L})}{[0.0821\text{ L} \cdot \text{atm}/(\text{mol} \cdot \text{K})](308\text{ K})} = 0.0811\text{ mol O}_2$$

Solved Problem 9.6 A 1.00 L sample of O_2 at 300 K and 1.00 atm plus a 0.500 L sample of N_2 at 300 K and 1.00 atm are put into a rigid 1.00 L container at 300 K. What will be their total volume, temperature, and pressure?

Solution: The total volume is the volume of the container, 1.00 L. The temperature is 300 K, given in the problem. The total pressure is the sum of the two partial pressures. The oxygen pressure is 1.00 atm. The nitrogen pressure is 0.500 atm, since it was moved from 0.500 L at 1.00 atm to 1.00 L at the same temperature (Boyle's law). The total pressure is

$$1.00 \text{ atm} + 0.500 \text{ atm} = 1.50 \text{ atm}$$

Solved Problem 9.7 O_2 is collected in a bottle over water at 25 °C at 1.00 atm barometric pressure. (*a*) What gas(es) is (are) in the bottle? (*b*) What is (are) the pressure(s)?

Solution: (*a*) Both O_2 and water vapor are in the bottle. (*b*) The total pressure is the barometric pressure, 760 torr. The water vapor pressure is 24 torr. The pressure of the O_2 must be

$$760 \text{ torr} - 24 \text{ torr} = 736 \text{ torr}$$

Solved Problem 9.8 Suppose that we double the length of each side of a rectangular box containing a gas. (*a*) What will happen to the volume? (*b*) What will happen to the pressure? (*c*) Explain the effect on the pressure on the basis of the kinetic molecular theory.

Solution: (*a*) The volume will increase by a factor of $(2)^3 = 8$. (*b*) The pressure will fall to one-eighth its original value. (*c*) In each direction, the molecules will hit the wall only one-half as often, and the force on each wall will drop to one-half of what it was originally because of this effect. Each wall has four times the area, and so the pressure will be reduced to one-fourth its original value because of this effect. The total reduction in pressure is $\frac{1}{2} \times \frac{1}{4} = \frac{1}{8}$, in agreement with Boyle's law.

Solved Problem 9.9 (*a*) If the velocity of a single gas molecule doubles, what happens to its kinetic energy? (*b*) If the average velocity of the molecules of a gas doubles, what happens to the temperature of the gas?

Solution: (*a*) $v_2 = 2\ v_1$

$$\text{KE}_2 = \frac{1}{2} m v_2^2 = \frac{1}{2} m (2v_1)^2 = 4\left(\frac{1}{2}\ m v_1^2\right) = 4\ \text{KE}_1$$

The kinetic energy is increased by a factor of four (*b*) The absolute temperature is increased by a factor of four.

Chapter 10
OXIDATION AND REDUCTION

IN THIS CHAPTER:

- ✔ *Assigning Oxidation Numbers*
- ✔ *Periodic Relationships of Oxidation Numbers*
- ✔ *Oxidation Numbers in Inorganic Nomenclature*
- ✔ *Balancing Oxidation-Reduction Equations*
- ✔ *Electrochemistry*
- ✔ *Solved Problems*

Assigning Oxidation Numbers

The term **oxidation** refers to a loss of electrons, while **reduction** means gain of electrons. Chemical reactions involving oxidation and reduction of atoms must be balanced not only in atoms but in electrons as well.

The **oxidation number** of an atom is defined as the number of valence electrons in the free atom minus the number "controlled" by the atom in the compound. If electrons are shared, "control" is given to

the more electronegative atom. For atoms of the same element, each atom is assigned one-half of the shared electrons. If electrons are transferred from one atom to another, the oxidation number equals the resulting charge. If electrons are shared, the oxidation number does not equal the charge; there may be no charge.

For example, consider CO_2.

	C	Each O
Number of valence electrons in free atom	4	6
–Number of valence electrons "controlled"	–0	–8
Oxidation number	+4	–2

Like the charge on an ion, each atom is assigned an oxidation number. The total of the oxidation numbers of all the atoms is equal to the net charge on the molecule or ion. Thus, for CO_2, the charge is $4 + 2(^-2) = 0$.

Learning these rules will facilitate the process of assigning oxidation numbers.

1. The sum of all the oxidation numbers in a species is equal to the charge on the species.
2. The oxidation number of uncombined elements is equal to 0.
3. The oxidation number of every monatomic ion is equal to its charge.
4. In its compounds, the oxidation number of every alkali metal and alkaline earth metal is equal to its group number.
5. The oxidation number of hydrogen in compounds is $^+1$ except when combined with active metals; then it is $^-1$.
6. The oxidation number of oxygen in its compounds is $^-2$ (with exceptions for peroxides and superoxides).
7. The oxidation number of every halogen atom in its compounds is $^-1$ except for a chlorine, bromine, or iodine atom combined with oxygen or a halogen atom higher in the periodic table.

Periodic Relationships of Oxidation Numbers

Oxidation numbers are very useful in correlating and systematizing a lot of inorganic chemistry. A few simple rules allow the prediction of the formulas of covalent compounds, just as predictions were made for ionic compounds in Chapter 4 by using the charges on the ions.

1. All elements when uncombined have oxidation numbers equal to 0. (Some also have oxidation numbers equal to 0 in some of their compounds).
2. The maximum oxidation number of any atom in any of its compounds is equal to its periodic group number, with a few exceptions. The coinage metals have the following maximum oxidation numbers: Cu, $^{+}2$; Ag, $^{+}2$; and Au, $^{+}3$. Some of the noble gases (group 0) have positive oxidation numbers. Some lanthanide and actinide element oxidation numbers exceed $^{+}3$, their nominal group number.
3. The minimum oxidation number of hydrogen is $^{-}1$. That of any other nonmetallic atom is equal to its group number minus 8. That of any metallic atom is 0.

Oxidation Numbers in Inorganic Nomenclature

In Chapter 5, Roman numerals were placed at the ends of names of metals to distinguish the charge on monatomic cations. This nomenclature is called the **Stock system**. It is really the oxidation number that is in parentheses. For monatomic ions, the oxidation number is equal to the charge. For other cations, again the oxidation number is used in the name. For example, Hg_2^{2+} is named mercury(I) ion. Its charge is 2^{+}; the oxidation number of each atom is $^{+}1$.

Balancing Oxidation-Reduction Equations

In every reaction in which the oxidation number of an element in one reactant (or more than one) increases, the oxidation number of another reactant (or more than one) must decrease. An increase in oxidation number is called oxidation; a decrease is reduction. The term **redox** is often used as a synonym for oxidation-reduction. The total change in oxidation number must be the same in the oxidation as in the reduction, because the number of electrons transferred from one species must be the same as the number transferred to the other. The species that causes another to be reduced is called the **reducing agent**; the species that causes the oxidation is called the **oxidizing agent**.

One important use of oxidation numbers is in balancing redox equations. There are essentially two methods to balance redox reactions: the **oxidation number change method** and the **ion-electron method**. In the former method, the changes in oxidation number are used to balance the species in which the elements that are oxidized and reduced appear. The numbers of atoms of each of these elements is used to give equal numbers of electrons gained and lost. If necessary, first balance the number of atoms of the element oxidized and/or the number of atoms of the element reduced. Then, balance by inspection, as was done in Chapter 7.

For example, balance the following equation:

$$? \text{ HCl} + ? \text{ HNO}_3 + ? \text{ CrCl}_2 \rightarrow ? \text{ CrCl}_3 + ? \text{ NO} + ? \text{ H}_2\text{O}$$

Inspect the oxidation states of all the elements. Notice that Cr goes from $^{+}2$ to $^{+}3$ (a change of $^{+}1$) and N goes from $^{+}5$ to $^{+}2$ (a change of $^{-}3$). To balance the oxidation numbers, three $CrCl_2$ and three $CrCl_3$ are needed for each N atom reduced.

$$? \text{ HCl} + 1 \text{ HNO}_3 + 3 \text{ CrCl}_2 \rightarrow 3 \text{ CrCl}_3 + 1 \text{ NO} + ? \text{ H}_2\text{O}$$

Next, balance the HCl by balancing the Cl atoms and balance the H_2O by balancing the O atoms.

$$3 \text{ HCl} + 1 \text{ HNO}_3 + 3 \text{ CrCl}_2 \rightarrow 3 \text{ CrCl}_3 + 1 \text{ NO} + 2 \text{ H}_2\text{O}$$

or

$$3 \text{ HCl} + \text{HNO}_3 + 3 \text{ CrCl}_2 \rightarrow 3 \text{ CrCl}_3 + \text{NO} + 2 \text{ H}_2\text{O}$$

In the ion-electron method of balancing redox equations, equations for the oxidation, and reduction half-reactions are written and balanced separately. Only when each of these is complete and balanced are the two combined into one complete equation for the reaction as a whole. In general, net ionic equations are used in this process. In the two half-reaction equations, electrons appear explicitly; in the complete reaction equation, no electrons are included.

One method of balancing redox equations by the half-reaction method is presented here. Steps one through five should be done for each half-reaction separately before proceeding to the rest of the steps.

1. Identify the element(s) oxidized and reduced. Write separate half-reactions for each of these.
2. Balance these elements.
3. Balance the change in oxidation number by adding electrons to the side with the higher total of oxidation numbers. That is, add electrons on the left for a reduction half-reaction and on the right for an oxidation half-reaction.
4. In acid solution, balance the net charge with hydrogen ions H^+. In basic solution, after all other steps have been completed, any H^+ can be neutralized by adding OH^- ions to each side, creating water and excess OH^- ions.
5. Balance the hydrogen and oxygen atoms with water.
6. If necessary, multiply every item in one or both sides of the equation by small integers so that the number of electrons is the same in each. The same small integer is used throughout each half-reaction and is different from that used in the other half-reaction. Then add the two half-reactions.
7. Cancel all species that appear on both sides of the equation. All the electrons must cancel out in this step, and often some hydrogen ions and water molecules also cancel.
8. Check to see that atoms of all the elements are balanced and that the net charge is the same on both sides of the equation.

To illustrate these steps, balance the following equation:

$$Fe^{2+} + H^+ + NO_3^- \rightarrow Fe^{3+} + NO + H_2O$$

Step 1:	$Fe^{2+} \rightarrow Fe^{3+}$	$NO_3^- \rightarrow NO$
Step 2:	Fe already balanced.	N already balanced.

Step 3: $Fe^{2+} \rightarrow Fe^{3+} + e^-$ $\qquad$ $3\,e^- + NO_3^- \rightarrow NO$

Step 4: $Fe^{2+} \rightarrow Fe^{3+} + e^-$ $\qquad$ $4\,H^+ + 3\,e^- + NO_3^- \rightarrow NO$

Step 5: $Fe^{2+} \rightarrow Fe^{3+} + e^-$ $\qquad$ $4\,H^+ + 3\,e^- + NO_3^- \rightarrow NO + 2\,H_2O$

Step 6: Multiplying by 3:

$3\,Fe^{2+} \rightarrow 3\,Fe^{3+} + 3\,e^-$

Adding:

$$3\,Fe^{2+} + 4\,H^+ + 3\,e^- + NO_3^- \rightarrow NO + 2\,H_2O + 3\,Fe^{3+} + 3\,e^-$$

Step 7: $3\,Fe^{2+} + 4\,H^+ + NO_3^- \rightarrow NO + 2\,H_2O + 3\,Fe^{3+}$

Step 8: The equation is balanced. There are 3 Fe atoms, 4 H atoms, 1 N atom, and 3 O atoms on each side. The net charge on each side is $^+9$.

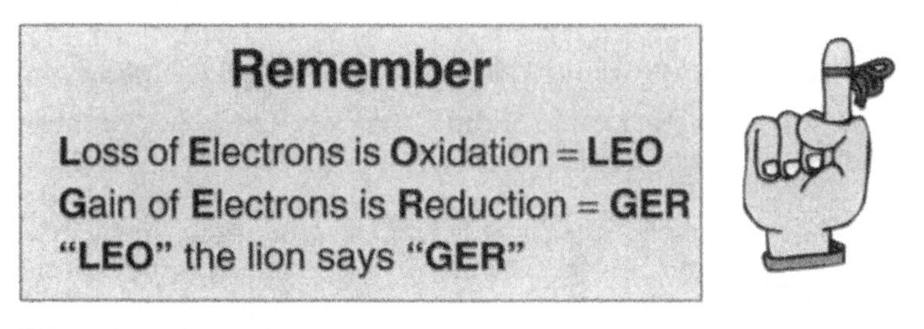

Electrochemistry

Oxidation reduction reactions occur at two electrodes. The electrode at which oxidation occurs is called the **anode**; the one at which reduction takes place is called the **cathode**. Electricity passes through a **circuit** under the influence of a **potential** or **voltage**, the driving force of the movement of charge. There are two different types of interaction of electricity and matter. **Electrolysis** is when an electric current causes a chemical reaction. **Galvanic cell action** is when a chemical reaction causes an electric current, as in the use of a battery.

Electrolysis. The requirements for electrolysis are as follows:

1. Ions to carry current.
2. Liquid, so that the ions can migrate.
3. Source of potential.
4. Mobile ions, complete circuit (including wires to carry electrons), and electrodes (at which the current changes from the flow of electrons to the movement of ions or vice versa).

The reaction conditions are very important to the products. If you electrolyze a solution containing a compound of a very active metal and/ or very active nonmetal, the water (or other solvent) might be electrolyzed instead of the ion. However, if you electrolyze a dilute aqueous solution of NaCl, the water is decomposed. The NaCl is necessary to conduct the current, but neither Na^+ nor Cl^- reacts at the electrodes. If you electrolyze a concentrated solution of NaCl instead, H_2 is produced at the cathode and Cl_2 is produced at the anode.

Electrolysis is used in a wide variety of ways. Electrolysis cells are used to produce very active elements in their elemental form. Electrolysis may be used to electroplate objects. Electrolysis is also used to purify copper, making it suitable to conduct electricity.

Galvanic Cells. When you place a piece of zinc metal into a solution of $CuSO_4$, you expect a chemical reaction because the more active zinc displaces the less active copper from its compound. This is a redox reaction, involving transfer of electrons from zinc to copper.

$$Zn \rightarrow Zn^{2+} + 2\,e^-$$
$$Cu^{2+} + 2\,e^- \rightarrow Cu$$

It is possible to carry out these same half-reactions in different places if connected suitably. The electrons must be delivered from Zn to Cu^{2+}, and there must be a complete circuit. The apparatus is shown in Figure 10-1. A galvanic cell with this particular combination of reactants is called a **Daniell cell**. The pieces of zinc and copper serve as electrodes, at which the electron current is changed to an ion current or vice versa. The salt bridge is necessary to complete the circuit. Electrons flow from the left to right through the wire, and they could be made to do electrical work, such as light a small bulb. To keep the beakers from acquiring a charge, cations flow through the salt bridge toward the right and anions flow to the left. The salt bridge is filled with a solution of an unreacting salt, such as KNO_3. The redox reaction provides the potential to produce the current in a complete circuit.

One such combination of anode and cathode is called a **cell**. Theoretically, any spontaneous redox reaction can be made to produce a galvanic cell. A combination of cells is called a **battery**.

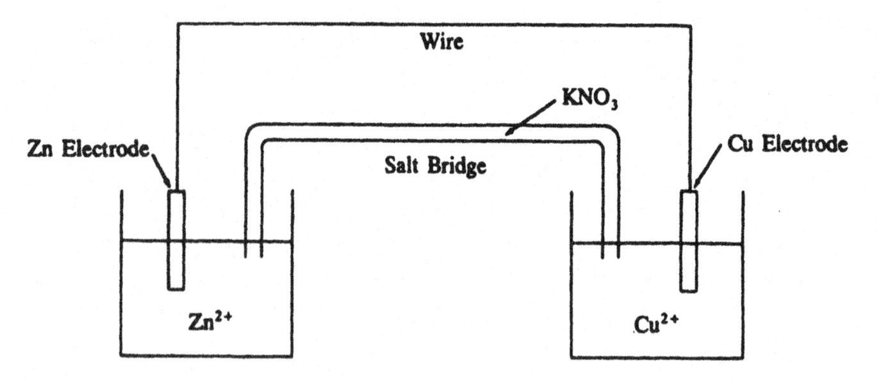

Figure 10-1 Daniell cell

Solved Problems

Solved Problem 10.1 Calculate the oxidation number of Cr in $Cr_2O_7^{2-}$.

Solution: There are two chromium atoms and seven oxygen atoms. The oxidation number of oxygen is $^-2$ and the total charge on the ion is $^-2$. Thus, the oxidation number of Cr is equal to x using the following equation:

$$2x + 7(^-2) = {}^-2$$
$$x = {}^+6$$

Solved Problem 10.2 Give the possible oxidation numbers for calcium.

Solution: Ca can have an oxidation number of 0 when it is a free element and $^+2$ in all its compounds.

Solved Problem 10.3 Name P_4O_{10} according to the Stock system.

Solution: Phosphorus(V) oxide.

Solved Problem 10.4 Balance the following equation using the oxidation number change method:

$$HCl + KMnO_4 + H_2C_2O_4 \rightarrow CO_2 + MnCl_2 + KCl + H_2O$$

Solution: Before attempting to balance the oxidation numbers gained and lost, balance the number of carbon atoms.

$$HCl + KMnO_4 + H_2C_2O_4 \rightarrow 2\ CO_2 + MnCl_2 + KCl + H_2O$$

Now proceed as before. Mn is reduced: $(^{+}7 \rightarrow {}^{+}2) = {}^{-}5$. C is oxidized: $2(^{+}3 \rightarrow {}^{+}4) = {}^{+}2$.

$$HCl + 2\ KMnO_4 + 5\ H_2C_2O_4 \rightarrow 10\ CO_2 + 2\ MnCl_2 + KCl + H_2O$$

Balance H_2O from the number of O atoms, KCl by the number of K atoms, and finally HCl by the number of Cl or H atoms. Check.

$$6\ HCl + 2\ KMnO_4 + 5\ H_2C_2O_4 \rightarrow 10\ CO_2 + 2\ MnCl_2 + 2\ KCl + 8\ H_2O$$

Solved Problem 10.5 Complete and balance the following equation in acid solution using the ion-electron method.

$$Cr_2O_7^{2-} + Cl_2 \rightarrow ClO_3^- + Cr^{3+}$$

Solution:

Step 1: $Cr_2O_7^{2-} \rightarrow Cr^{3+}$ | $Cl_2 \rightarrow ClO_3^-$

Step 2: $Cr_2O_7^{2-} \rightarrow 2\ Cr^{3+}$ | $Cl_2 \rightarrow 2\ ClO_3^-$

Step 3: 2 atoms are reduced 3 units each. | 2 atoms are oxidized 5 units each.

$6\ e^- + Cr_2O_7^{2-} \rightarrow 2\ Cr^{3+}$ | $Cl_2 \rightarrow 2\ ClO_3^- + 10\ e^-$

Step 4: $14\ H^+ + 6\ e^- + Cr_2O_7^{2-} \rightarrow 2\ Cr^{3+}$ | $Cl_2 \rightarrow 2\ ClO_3^- + 10\ e^- + 12\ H^+$

Step 5: $14\ H^+ + 6\ e^- + Cr_2O_7^{2-} \rightarrow 2\ Cr^{3+} + 7\ H_2O$ | $6\ H_2O + Cl_2 \rightarrow 2\ ClO_3^- + 10\ e^- + 12\ H^+$

Step 6: $5\ [14\ H^+ + 6\ e^- + Cr_2O_7^{2-} \rightarrow 2\ Cr^{3+} + 7\ H_2O]$ | $3\ [6\ H_2O + Cl_2 \rightarrow 2\ ClO_3^- + 10\ e^- + 12\ H^+]$

$70\ H^+ + 30\ e^- + 5\ Cr_2O_7^{2-} \rightarrow 10\ Cr^{3+} + 35\ H_2O$ | $18\ H_2O + 3\ Cl_2 \rightarrow 6\ ClO_3^- + 30\ e^- + 36\ H^+]$

Step 7:

$$18\ H_2O + 3\ Cl_2 + 70\ H^+ + 30\ e^- + 5\ Cr_2O_7^{2-} \rightarrow 10\ Cr^{3+} + 35\ H_2O + 6\ ClO_3^- + 30\ e^- + 36\ H^+$$

$$3\ Cl_2 + 34\ H^+ + 5\ Cr_2O_7^{2-} \rightarrow 10\ Cr^{3+} + 17\ H_2O + 6\ ClO_3^-$$

Step 8: The equation is balanced. There are 6 Cl atoms, 34 H atoms, 10 Cr atoms, and 35 O atoms on each side, as well as a net $^{+}24$ charge on each side.

Solved Problem 10.6 The electrolysis of brine (concentrated NaCl solution) produces hydrogen at the cathode and chlorine at the anode. Write a net ionic equation for each half-reaction and the total reaction.

Solution:

$$2\,Cl^- \rightarrow Cl_2 + 2\,e^-$$
$$2\,H_2O + 2\,e^- \rightarrow H_2 + 2\,OH^-$$
$$2\,Cl^- + 2\,H_2O \rightarrow Cl_2 + H_2 + 2\,OH^-$$

Chapter 11
SOLUTIONS

IN THIS CHAPTER:

- ✔ *Qualitative Concentration Terms*
- ✔ *Molarity*
- ✔ *Titration*
- ✔ *Molality*
- ✔ *Mole Fraction*
- ✔ *Equivalents*
- ✔ *Normality*
- ✔ *Solved Problems*

Qualitative Concentration Terms

Solutions are mixtures and therefore do not have definite compositions. For most solutions there is a limit to how much **solute** will dissolve in a given quantity of **solvent** at a given temperature. The maximum concentration of solute that will dissolve in contact with solute is called **solubility** of the solute. Solubility depends on temperature. Most solids dissolve in liquids more readily at high temperatures than at low temperatures, while gases dissolve in cold liquids better than in hot liquids.

A solution in which the concentration of the solute is equal to the solubility is called a **saturated** solution. If the concentration is lower, the solution is said to be **unsaturated**. It is also possible to prepare a **supersaturated** solution, an unstable solution containing a greater concentration of solute than is present in a saturated solution. Such a solution deposits the excess solute if a crystal of the solute is added to it. It is prepared by dissolving solute at one temperature and carefully changing the temperature to a point where the solution is unstable.

Molarity

Perhaps the most useful measure of concentration is **molarity**. Molarity is defined as the number of moles of solute per liter of solution. It is abbreviated **M**. Being a ratio, molarity can be used as a factor.

When water is added to a solution, the volume increases, but the number of total moles present does not change. The molarity of every solute in the solution therefore decreases.

Titration

To determine the concentration of a solution, a solution of known concentration and volume can be treated with the unknown solution until the mole ratio is exactly what is required by the balanced chemical equation. Then from the known volumes of both reactants, the concentration of the unknown can be calculated. This procedure is called **titration**. An **indicator** is used to tell when to stop the titration. Typically an indicator is a compound that is one color (or colorless) in an acidic solution and a second color in a basic solution.

In a typical titration experiment, 25.00 mL of 2.000 M HCl is pipetted into an Erlenmeyer flask. A **pipet** is a piece of glassware that is calibrated to deliver an exact volume of liquid. A solution of NaOH of unknown concentration is placed in a **buret** (see Figure 11-1), and some is allowed to drain out of the bottom to ensure that the portion below the stopcock is filled. The buret volume is read before any NaOH is added from it to the HCl (say, 3.30 mL) and again after the NaOH is added (say, 45.32 mL). The volume of added NaOH is merely the difference in readings (45.32 mL − 3.30 mL = 42.02 mL). The concentration of NaOH may

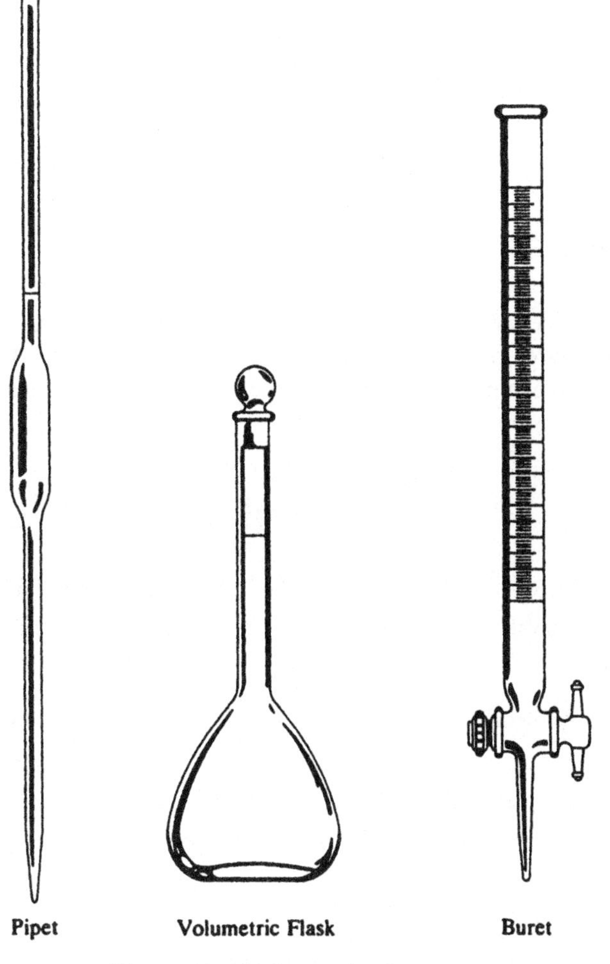

Figure 11-1 Volumetric glassware

now be calculated because the exact number of moles of NaOH has been added to exactly react with HCl.

$$25.00 \text{ mL HCl}\left(\frac{2.000 \text{ mmol HCl}}{1 \text{ mL HCl}}\right) = 50.00 \text{ mmol HCl}$$

The number of millimoles of NaOH is exactly the same, since the addition was stopped when the reaction was just complete. Therefore, there is 50.00 mmol NaOH in the volume of NaOH added,

$$\frac{50.00 \text{ mmol NaOH}}{42.02 \text{ mL}} = 1.190 \text{ M NaOH}$$

Thus, the concentration of the unknown NaOH was 1.190 M.

Molality

Molarity is defined in terms of the volume of solution. Since the volume is temperature-dependent, so is the molarity of the solution. **Molality** is defined as the number of moles of solute per kilogram of solvent in a solution. The symbol for molality is **m**.

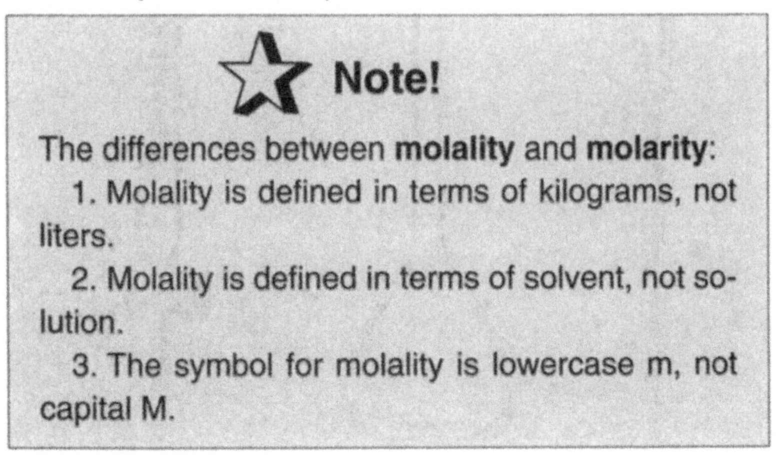

Note!

The differences between **molality** and **molarity**:

1. Molality is defined in terms of kilograms, not liters.
2. Molality is defined in terms of solvent, not solution.
3. The symbol for molality is lowercase m, not capital M.

Mole Fraction

The **mole fraction** of a substance in a solution is the ratio of the number of moles of that substance to the total number of moles in the solution. The symbol for mole fraction is usually X_A. Thus, for a solution containing x mol of A, y mol of B, and z mol of C, the mole fraction of A is

$$X_A = \frac{x \text{ mol A}}{x \text{ mol A} + y \text{ mol B} + z \text{ mol C}}$$

Equivalents

Equivalents are measures of the quantity of a substance present, analogous to moles. The equivalent is defined in terms of a chemical reaction. It is defined in one of two different ways, depending on whether a redox reaction or an acid-base reaction is under discussion. For a redox reaction, one equivalent is the quantity of a substance that will react with or yield one mol of electrons. For an acid-base reaction, one equivalent is the quantity of a substance that will react with or yield one mol of hydrogen ions or hydroxide ions.

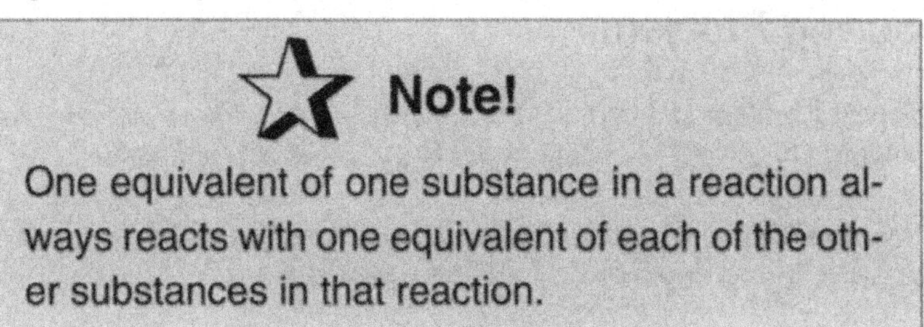

The major use of equivalents stems from its definition. Once you define the number of equivalents in a certain mass of a substance, you do not need to write the equation for its reaction. That equation has already been used in defining the number of equivalents. Thus, a chemist can calculate the number of equivalents in a certain mass of substance, and technicians can subsequently use that definition without knowing the details of the reaction. The **equivalent mass** of a substance is the mass in grams of one equivalent of the substance.

Normality

Analogous to molarity, **normality** is defined as the number of equivalents of solute per liter of solution. Its unit is **normal** with the symbol **N**. Normality is some integral multiple of molarity, since there are always some integral number of equivalents per mole. Since one equivalent of one thing reacts with one equivalent of any other thing in the reaction, it is also true that the volume times the normality of the first thing is equal to the volume times the normality of the second.

$$N_1V_1 = N_2V_2$$

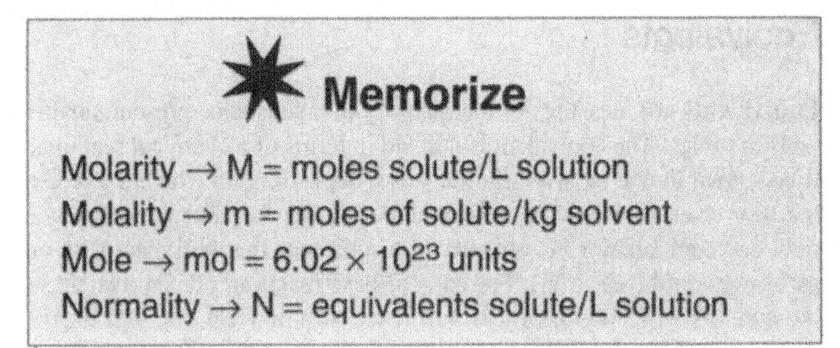

Solved Problems

Solved Problem 11.1 What is the molarity of a solution prepared by dissolving 117 g NaCl in enough water to make 500 mL of solution?

Solution: Molarity is defined in moles per liter. We can convert each of the quantities given to those used in the definition:

$$117\text{ g NaCl}\left(\frac{1\text{ mol NaCl}}{58.5\text{ g NaCl}}\right) = 2.00\text{ mol NaCl}$$

$$500\text{ mL}\left(\frac{1\text{ L}}{1000\text{ mL}}\right) = 0.500\text{ L}$$

$$\text{Molarity} = \frac{2.00\text{ mol}}{0.500\text{ L}} = 4.00\text{ M}$$

Solved Problem 11.2 Calculate the final concentration if 2.0 L of 3.0 M NaCl and 4.0 L of 1.5 M NaCl are mixed.

Solution:

$$\text{Final volume} = 6.0\text{ L}$$

$$\text{Final moles} = 2.0\text{ L}\left(\frac{3.0\text{ mol}}{1\text{ L}}\right) + 4.0\text{ L}\left(\frac{1.5\text{ mol}}{1\text{ L}}\right) = 12.0\text{ mol}$$

$$\text{Molarity} = \frac{12.0\text{ mol}}{6.0\text{ L}} = 2.0\text{ M}$$

Solved Problem 11.3 What is the concentration of a $Ba(OH)_2$ solution if it takes 47.70 mL to neutralize 25.00 mL of 1.444 M HCl?

Solution: The number of moles of HCl is easily calculated.

$$0.02500 \text{ L}\left(\frac{1.444 \text{ mol}}{1 \text{ L}}\right) = 0.03610 \text{ mol HCl}$$

The balanced chemical equation shows that the ratio of moles of HCl to $Ba(OH)_2$ is 2:1.

$$2 \text{ HCl} + Ba(OH)_2 \rightarrow BaCl_2 + 2\, H_2O$$

$$0.3610 \text{ mol HCl}\left[\frac{1 \text{ mol } Ba(OH)_2}{2 \text{ mol HCl}}\right] = 0.01805 \text{ mol } Ba(OH)_2$$

The molarity is given by

$$\frac{0.01805 \text{ mol } Ba(OH)_2}{0.04770 \text{ L}} = 0.3784 \text{ M } Ba(OH)_2$$

Solved Problem 11.4 What mass of 3.00 m CH_3OH solution in water can be prepared with 50.0 g of CH_3OH?

Solution: Keep careful track of the units.

$$50.0 \text{ g } CH_3OH\left(\frac{1 \text{ mol } CH_3OH}{32.0 \text{ g } CH_3OH}\right)\left(\frac{1.00 \text{ kg solvent}}{3.00 \text{ mol } CH_3OH}\right)$$
$$= 0.521 \text{ kg solvent}$$

$$50.0 \text{ g } CH_3OH + 521 \text{ g } H_2O = 571 \text{ g solution}$$

Solved Problem 11.5 What is the mole fraction of CH_3OH in a solution of 10.0 g CH_3OH and 20.0 g H_2O?

Solution:

$$10.0\ \text{g CH}_3\text{OH}\left(\frac{1\ \text{mol CH}_3\text{OH}}{32.0\ \text{g CH}_3\text{OH}}\right) = 0.313\ \text{mol CH}_3\text{OH}$$

$$20.0\ \text{g H}_2\text{O}\left(\frac{1\ \text{mol H}_2\text{O}}{18.0\ \text{g H}_2\text{O}}\right) = 1.11\ \text{mol H}_2\text{O}$$

$$X_{\text{CH}_3\text{OH}} = \frac{0.313\ \text{mol CH}_3\text{OH}}{0.313\ \text{mol CH}_3\text{OH} + 1.11\ \text{mol H}_2\text{O}} = 0.220$$

Since the mole fraction is a ratio of moles (of one substance) to moles (total), the units cancel and mole fraction has no units.

Solved Problem 11.6 How many equivalents are there in 98.0 g of H_2SO_4 in the following reaction?

$$\text{H}_2\text{SO}_4 + \text{NaOH} \rightarrow \text{NaHSO}_4 + \text{H}_2\text{O}$$

Solution: This is an acid-base reaction, so the number of equivalents of H_2SO_4 is defined in terms of the number of moles of hydroxide ion with which it reacts. According to the equation, 1 mol of H_2SO_4 reacts with 1 mol of OH^- so 1 equivalent of H_2SO_4 is equal to 1 mol of H_2SO_4. Since 98.0 g H_2SO_4 is 1 mol, there is 1 equivalent in 98.0 g H_2SO_4.

Solved Problem 11.7 How many equivalents are there in 98.0 g of H_2SO_4 in the following reaction?

$$8\ \text{H}^+ + \text{H}_2\text{SO}_4 + 4\ \text{Zn} \rightarrow \text{H}_2\text{S} + 4\ \text{Zn}^{2+} + 4\ \text{H}_2\text{O}$$

Solution: This is a redox reaction, and so the number of equivalents of H_2SO_4 is defined in terms of the number of moles of electrons with which it reacts. Since no electrons appear explicitly in an overall equation, we will write the half-reaction in which the H_2SO_4 appears:

$$8\ \text{H}^+ + \text{H}_2\text{SO}_4 + 8\ e^- \rightarrow \text{H}_2\text{S} + 4\ \text{H}_2\text{O}$$

It is now apparent that 1 mol of H_2SO_4 reacts with 8 mol e^-, and by definition 8 equivalents of H_2SO_4 react with 8 mol e^-. Thus, 8 equivalents equals 1 mol in this reaction. Since 98.0 g H_2SO_4 is 1 mol, there are 8 equivalents in 98.0 g H_2SO_4.

Solved Problem 11.8 What volume of 3.00 N NaOH is required to neutralize 25.00 mL of 3.50 N H_2SO_4?

Solution:

$$N_1V_1 = N_2V_2$$

$$V_2 = \frac{N_1V_1}{N_2} = \frac{(3.50\text{ N H}_2\text{SO}_4)(25.00\text{ mL})}{3.00\text{ N NaOH}} = 29.2\text{ mL}$$

Note that less volume of H_2SO_4 is required than of NaOH because its normality is greater.

Chapter 12
RATES AND EQUILIBRIUM

IN THIS CHAPTER:

- ✔ *Collision Theory*
- ✔ *Rates of Chemical Reaction*
- ✔ *Chemical Equilibrium*
- ✔ *Equilibrium Constants*
- ✔ *Solved Problems*

Collision Theory

The collision theory examines the molecules undergoing reaction to explain the observed phenomena. The theory postulates that in order for a reaction to occur, molecules must collide with one another with sufficient energy to break chemical bonds in the reactants. A very energetic and highly unstable species is formed, called an **activated complex**. Not every collision between reacting molecules, even those with sufficient energy, produces products. The molecules might be oriented in the wrong directions, or the activated complex may break up to reform the reactants instead of forming the products. But the huge majority of collisions do not have enough energy to cause bond breakage in the first place.

The minimum energy that may cause a reaction to occur is called the **activation energy**, designated E_a (see Figure 12-1). If molecular colli-

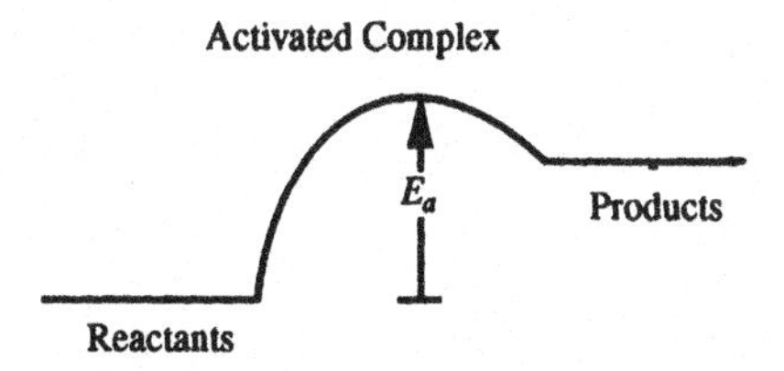

Figure 12-1 Activation energy

sions are not energetic enough, the molecules will merely return to their original states even if temporarily they have been somewhat deformed.

Rates of Chemical Reaction

The **rate of a reaction** is defined as the change in concentration of any of its reactants or products per unit time. There are six factors that affect rate of a reaction:

1. The nature of the reactants. This factor is least controllable by the chemist.
2. Temperature. In general, the higher the temperature of the system, the faster the chemical reaction will proceed.
3. The presence of a catalyst. A catalyst is a substance that can accelerate a chemical reaction without undergoing a permanent change in its own composition.
4. The concentration of the reactants. In general, the higher the concentration of the reactants, the faster the reaction.
5. The pressure of the gaseous reactants. In general, the higher the pressure, the faster the reaction.
6. State of subdivision. The smaller the pieces of a solid reactant, the smaller state of subdivision, the faster the reaction.

The collision theory allows us to explain the factors that affect reaction rate. There is a wide range of energies among molecules in any sample, and generally only the most energetic molecules can undergo reaction. An increase in temperature increases the number of molecules that have sufficient energy to react. An increase in concentration or pressure causes the molecules to collide more often, and with more collisions, more effective collisions are expected. The state of subdivision affects the reaction rate because the more surface area there is, the more collisions there are between the fluid molecules and the solid surface. A catalyst works by reducing the activation energy, making an easier path for the reactants to get to products. Since more reactant molecules have this lower activation energy, the reaction goes faster.

You Need to Know ✓

A rise in temperature of 10 °C approximately doubles the reaction rate.

Chemical Equilibrium

Many chemical reactions convert practically all the reactant(s) to products under a given set of conditions. These reactions are said to **go to completion**. In other reactions, as the products are formed, they in turn react to form the original reactants again. This situation, two opposing reactions occurring at the same time, leads to the formation of some products, but the reactants are not completely converted to products. A state in which two exactly opposite reactions are occurring at the same rate is called **chemical equilibrium**. For example, nitrogen and hydrogen gases react with each other at 500 °C and high pressure to form ammonia; under the same conditions, ammonia decomposes to produce hydrogen and nitrogen:

$$3\ H_2 + N_2 \rightarrow 2\ NH_3$$
$$2\ NH_3 \rightarrow 3\ H_2 + N_2$$

These two exactly opposite equations may be written as one, with double arrows:

$$3\ H_2 + N_2 \rightleftarrows 2\ NH_3 \quad \text{or} \quad 2\ NH_3 \rightleftarrows 3\ H_2 + N_2$$

The reagents written on the right of the chemical reaction are the **products** and those on the left are the **reactants**, despite the fact that the equation may be written with either set of reagents on either side.

When the hydrogen and nitrogen are placed in a vessel together, at first there is no ammonia present, and so the only reaction that occurs is the combination of the two elements. As time passes, there is less and less nitrogen and hydrogen, and the combination reaction therefore slows down. Meanwhile, the concentration of ammonia is building up, and the decomposition rate of the ammonia therefore increases. There comes a time when both the combination and the decomposition reaction occur at the same rate. When that happens, the concentration of ammonia will not change anymore. The reaction apparently stops. However, in reality, both reactions continue to occur; their effects merely cancel each other out. A state of equilibrium has been reached.

If the conditions on an equilibrium system are changed, some further reaction may be achieved. Soon, however, the system will achieve a new equilibrium at the new set of conditions. **Le Châtelier's principle** states that if stress is applied to a system at equilibrium, the equilibrium will shift in a tendency to reduce that stress. A **stress** is something done to the system. For example, an increase in concentration of one of the reactants or products will cause the equilibrium to shift to try to reduce that concentration increase. If heat is added to a system, the equilibrium will react to reduce that stress by using up some of the added heat. An increase in pressure will be compensated by a shift to reduce the total number of moles present. The addition of a catalyst will not cause any change in the position of the equilibrium; it will shift neither left nor right. However, it will speed up both the forward and reverse reaction rates equally.

Le Châtelier's Principle: If an equilibrium system is stressed, a reaction will occur in the direction which tends to relieve the stress.

Equilibrium Constants

Although Le Châtelier's principle does not tell us how much an equilibrium will be shifted, there is a way to determine the position of an equilibrium once data have been determined for the equilibrium experimentally. At a given temperature the ratio of concentrations of products to reactants, each raised to a suitable power, is constant for a given equilibrium reaction. The letters A, B, C, and D are used here to stand for general chemical species. Thus, for a chemical reaction in general,

$$aA + bB \rightleftarrows cC + dD$$

The following ratio always has the same value at a given temperature:

$$K = \frac{[C]^c[D]^d}{[A]^a[B]^b}$$

Here the square brackets indicate the concentration of the chemical species within the bracket. That is, [A] means the concentration of A, and so forth. $[A]^a$ means the concentration of A raised to the power a, where a is the value of the coefficient of A in the balanced equation for the chemical equilibrium. The value of the ratio of concentration terms is symbolized by the letter ***K***, called the **equilibrium constant**.

Solved Problems

Solved Problem 12.1 If the activation energy of a certain reaction is 25 kJ/mol and the overall reaction process produces 15 kJ/mol of energy in going from reactants to products, how much energy is given off when the activated complex is converted to products?

Solution: The process produces 40 kJ/mol:

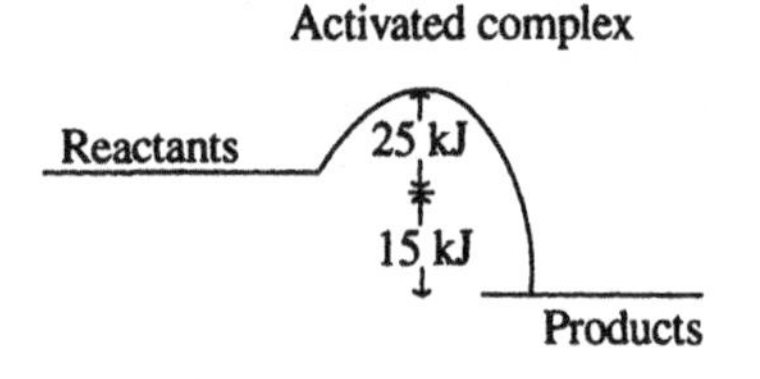

Solved Problem 12.2 For the reaction

$$A + B \rightleftarrows C + \text{heat}$$

does a rise in temperature increase or decrease the rate of (*a*) the forward reaction and (*b*) the reverse reaction? (*c*) Which effect is greater?

Solution: (*a*) Increase and (*b*) increase. (An increase in temperature increases all rates.) (*c*) The added heat shifts the equilibrium to the left. That means the reverse reaction rate was increased more than the forward reaction rate.

Solved Problem 12.3 What effect would a decrease in volume have on the following system at equilibrium at 500 °C?

$$2\ C(s) + O_2 \rightleftarrows 2\ CO$$

Solution: The equilibrium would shift to the left. The decrease in volume would increase the pressure of each of the gases, but not of the carbon, which is a solid. The number of moles of gas would be decreased by the shift to the left.

Solved Problem 12.4 Calculate the value of the equilibrium constant for the following reaction if at equilibrium there are 1.10 mol A, 1.60 mol B, 0.250 mol C, and 0.750 mol D in 500 mL of solution.

$$A + B \rightleftarrows C + D$$

Solution: Since number of moles and a volume are given, it is easy to calculate the equilibrium concentrations of the species. In this case, [A] = 2.20 M, [B] = 3.20 M, [C] = 0.500 M, and [D] = 1.50 M. Substitute the concentrations into the equilibrium constant expression. Note that all the coefficients are equal to 1. Solve.

$$K = \frac{[C]\,[D]}{[A]\,[B]}$$

$$K = \frac{(0.500\text{ M})\,(1.50\text{ M})}{(2.20\text{ M})(3.20\text{ M})} = 0.107$$

Solved Problem 12.5 Calculate the value of the equilibrium constant in the following reaction if 1.00 mol of A and 2.00 mol of B are placed in 1.00 L of solution and allowed to come to equilibrium. The equilibrium concentration of C is found to be 0.30 M.

$$A + B \rightleftarrows C + D$$

Solution: To determine the equilibrium concentrations of all the reactants and products, deduce the changes that have occurred. Assume some A and B have been used up, and that some D has also been produced. It is perhaps easiest to tabulate the various concentrations. Use the chemical equation as the table headings and enter the known values.

	A	+	B	⇄	C	+	D
Initial [ ]	1.00		2.00		0.00		0.00
Changes							
Equilibrium [ ]					0.30		

To produce 0.30 M C it takes 0.30 M A and 0.30 M B. Moreover, 0.30 M D was also produced. The magnitudes of the values in the second row of the table, the changes produced by the reaction, are always in the same ratio as the coefficients in the balanced chemical equation.

	A	+	B	⇄	C	+	D
Initial [ ]	1.00		2.00		0.00		0.00
Changes	−0.30		−0.30		+0.30		+0.30
Equilibrium [ ]					0.30		

Add the columns to find the rest of the equilibrium concentrations.

	A	+	B	⇄	C	+	D
Initial [ ]	1.00		2.00		0.00		0.00
Changes	−0.30		−0.30		+0.30		+0.30
Equilibrium [ ]	0.70		1.70		0.30		0.30

Now, substitute the equilibrium concentration values into the equilibrium constant expression.

$$K = \frac{[C]\,[D]}{[A]\,[B]} = \frac{(0.30)(0.30)}{(0.70)(1.70)} = 0.076$$

Solved Problem 12.6 In the reaction

$$X + Y \rightleftarrows Z$$

1 mol of X, 1.7 mol of Y, and 2.4 mol of Z are found at equilibrium in a 1.0 L reaction mixture. (*a*) Calculate *K*. (*b*) If the same mixture had been found in a 2.0 L reaction mixture, would the value of *K* have been the same? Explain.

Solution:

$$(a)\ K = \frac{[Z]}{[X][Y]} = \frac{2.4}{(1.0)(1.7)} = 1.4$$

$$(b)\ K = \frac{1.2}{(0.50)(0.85)} = 2.8$$

The value is not the same. The value of *K* is related to the concentrations of the reagents, not to their number of moles.

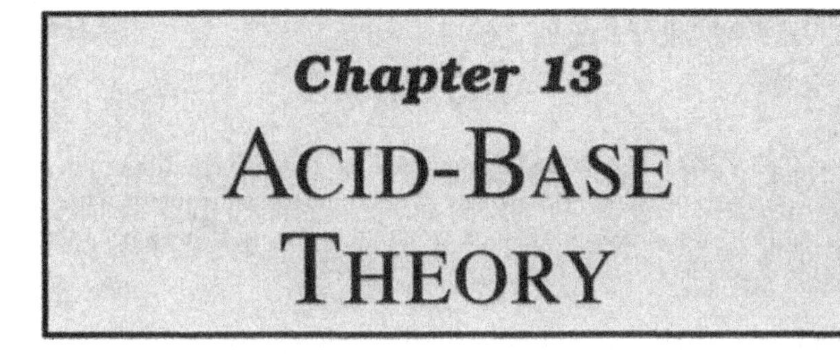

IN THIS CHAPTER:

- ✔ *The Brønsted Theory*
- ✔ *Acid-Base Equilibrium*
- ✔ *Autoionization of Water*
- ✔ *pH*
- ✔ *Buffer Solutions*
- ✔ *Solved Problems*

The Brønsted Theory

Thus far, only the **Arrhenius theory** of acids and bases has been considered. Acids are defined as hydrogen-containing compounds that react with bases. Bases are compounds containing OH^- ions or that form OH^- when they react with water. Bases react with acids to form salts and water.

The **Brønsted theory** expands these definitions of acids and bases to explain much more of solution chemistry. For example, the Brønsted theory explains why a solution of ammonium chloride tests acidic and a solution of sodium acetate tests basic. In the Brønsted theory, an **acid** is defined as a substance that donates a proton to another substance. In this sense,

a proton is a hydrogen atom that has lost its electron. A base is a substance that accepts a proton from another substance. The reaction of an acid and a base produces another acid and base. The following reaction is thus an acid-base reaction according to Brønsted:

$$HC_2H_3O_2 + H_2O \rightleftarrows C_2H_3O_2^- + H_3O^+$$

The $HC_2H_3O_2$ is an acid because it donates its proton to the H_2O to form $C_2H_3O_2^-$ and H_3O^+. The H_2O is a base because it accepts that proton. But this is an equilibrium reaction, and $C_2H_3O_2^-$ reacts with H_3O^+ to form $HC_2H_3O_2$ and H_2O. The $C_2H_3O_2^-$ is a base because it accepts the proton from H_3O^+; the H_3O^+ is an acid because it donates a proton.

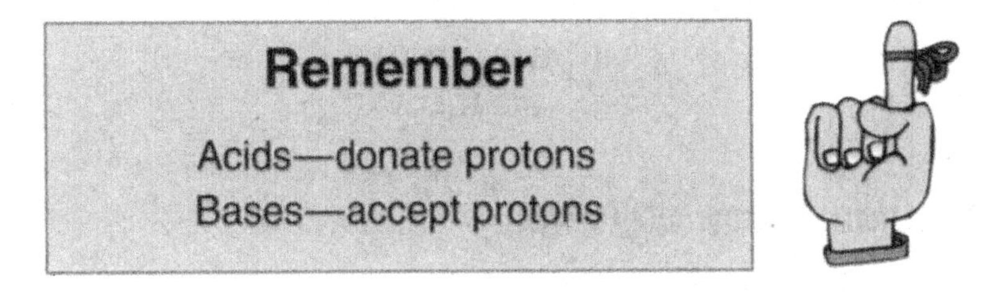

The acid on the left of this equation is related to the base on the right; they are said to be **conjugates** of each other. The $HC_2H_3O_2$ is the conjugate acid of the base $C_2H_3O_2^-$. Similarly, H_2O is the conjugate base of H_3O^+. The stronger the acid, the weaker its conjugate base.

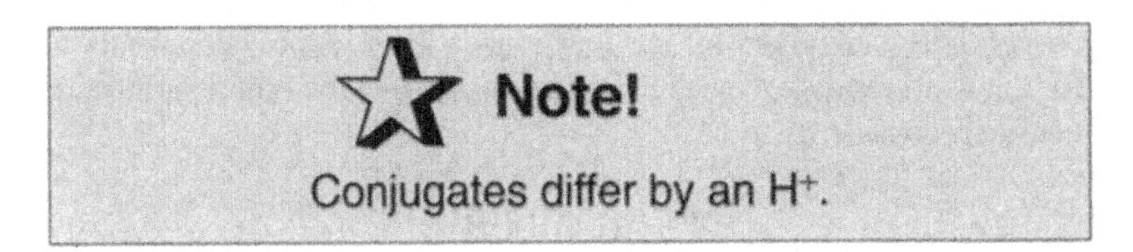

Water can serve as either an acid or a base. It is said to be **amphiprotic** because it reacts as an acid in the presence of bases, and it reacts as a base in the presence of acids.

Acid-Base Equilibrium

Equilibrium constants can be written for the ionization of weak acids and weak bases, just as for any other equilibria. For the equation

$$HC_2H_3O_2 + H_2O \rightleftarrows C_2H_3O_2^- + H_3O^+$$

we would originally (Chapter 12) write

$$K = \frac{[C_2H_3O_2^-][H_3O^+]}{[HC_2H_3O_2][H_2O]}$$

However, in dilute aqueous solution, the concentration of H_2O is practically constant, and its concentration is conventionally built into the value of the equilibrium constant. The new constant, called K_a or K_i for acids (K_b or K_i for bases), does not have the water concentration term in the denominator:

$$K_a = \frac{[C_2H_3O_2^-][H_3O^+]}{[HC_2H_3O_2]}$$

Autoionization of Water

Since water is defined as both an acid and a base, it is not surprising to find that water can react with itself, even though only to a very limited extent, in a reaction alled **autoionization**:

$$H_2O + H_2O \rightleftarrows H_3O^+ + OH^-$$

An equilibrium constant for this reaction, called K_w, does not have terms for the concentration of water; otherwise it is like the other equilibrium constants considered so far.

$$K_w = [H_3O^+][OH^-]$$

The value for this constant in dilute aqueous solution at 25 °C is 1.0×10^{-14}. Thus, water ionizes very little.

The equation for K_w means that there is always some H_3O^+ and always some OH^- in any aqueous solution. Their concentrations are inversely proportional. A solution is acidic if the H_3O^+ concentration exceeds the OH^- concentration; it is neutral if the two concentrations are equal; and it is basic if the OH^- concentration exceeds the H_3O^+ concentration.

pH

The pH scale was invented to reduce the necessity for using exponential numbers to report acidity. The pH is defined as

$$pH = -\log[H_3O^+]$$

From the way pH is defined and the value of K_w, we can deduce that solutions with pH = 7 are neutral, those with pH less than 7 are acidic, and those with pH greater than 7 are basic.

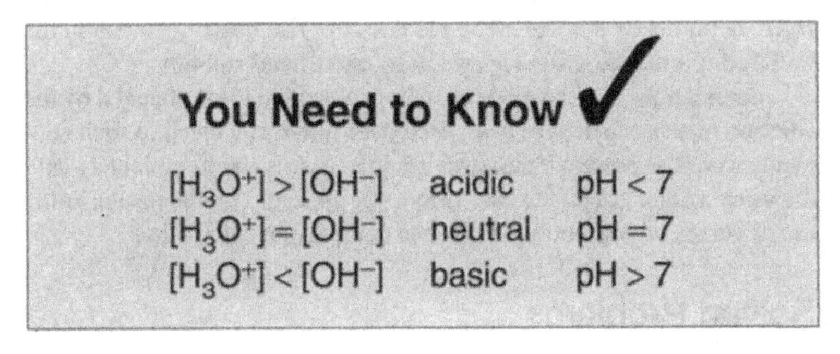

Buffer Solutions

A **buffer solution** is a solution of a weak acid and its conjugate base or a weak base and its conjugate acid. The main property of a buffer solution is its resistance to changes in its pH despite the addition of small quantities of strong acid or strong base. A buffer solution may be prepared by the addition of a weak acid to a salt of that acid or addition of a weak base to a salt of that base. For example, a solution of acetic acid and sodium acetate is a buffer solution.

A buffer solution resists change in its acidity. For example, a certain solution of acetic acid and sodium acetate has a pH of 4.0. When a small quantity of NaOH is added, the pH goes up to 4.2. If that quantity of NaOH had been added to that same volume of an unbuffered solution of HCl at pH 4, the pH would have gone up to 12.

The buffer solution works on the basis of Le Châtelier's principle. Consider the equation for the reaction of acetic acid with water:

$$HC_2H_3O_2 + H_2O \rightleftarrows C_2H_3O_2^- + H_3O^+$$

The solution of $HC_2H_3O_2$ and $C_2H_3O_2^-$ in H_2O results in the following relative quantities: $HC_2H_3O_2$ and $C_2H_3O_2^-$ are in excess, H_2O is in huge excess, and H_3O^+ is in limited quantity. If H_3O^+ is added to the equilibrium system, the equilibrium shifts to use up some of the added H_3O^+. If the acetate ion were not present to take up the added H_3O^+, the pH would drop. Since the acetate ion reacts with much of the added H_3O^+, there is little increase in H_3O^+ and little drop in pH. If OH^- is added to the solution, it reacts with H_3O^+ present. But the removal of that H_3O^+ is a stress, which causes this equilibrium to shift to the right, replacing much of the H_3O^+ removed by the OH^-. The pH does not rise nearly as much in the buffered solution as it would have in an unbuffered solution.

Calculations may be made as to how much the pH is changed by the addition of a strong acid or base. First determine how much of each conjugate would be present if that strong acid or base reacted completely with the weak acid or conjugate base originally present. Use the results as the initial values of concentrations for the equilibrium calculations.

Solved Problems

Solved Problem 13.1 Write an equilibrium equation for the reaction of NH_3 and H_2O. Indicate the conjugate acids and bases.

Solution:

$$NH_3 + H_2O \rightleftarrows NH_4^+ + OH^-$$

NH_3 is a base because it accepts a proton from water, which is therefore an acid. The NH_4^+ is an acid because it can donate a proton to OH^-, a base. NH_3 and NH_4^+ are conjugates. H_2O and OH^- are conjugates.

Solved Problem 13.2 The $[H_3O^+]$ in 0.250 M benzoic acid ($HC_7H_5O_2$) is 3.9×10^{-3} M. Calculate K_a.

Solution:

$$HC_7H_5O_2 + H_2O \rightleftarrows H_3O^+ + C_7H_5O_2^-$$

$$[H_3O^+] = [C_7H_5O_2^-] = 3.9 \times 10^{-3}$$
$$[HC_7H_5O_2] = 0.250 - (3.9 \times 10^{-3}) = 0.246$$
$$K_a = \frac{[H_3O^+][C_7H_5O_2^-]}{[HC_7H_5O_2]} = \frac{(3.9 \times 10^{-3})^2}{0.246} = 6.2 \times 10^{-5}$$

Solved Problem 13.3 Calculate the hydronium ion concentration in 0.010 M NaOH.

Solution:

$$K_w = [H_3O^+]\,[OH^-] = 1.0 \times 10^{-14}$$
$$[OH^-] = 0.010 \text{ M}$$
$$[H_3O^+]\,(0.010) = 1.0 \times 10^{-14}$$
$$[H_3O^+] = 1.0 \times 10^{-12}$$

Solved Problem 13.4 Calculate the pH of 0.10 M HCl.

Solution:

$$[H_3O^+] = 0.10 = 1.0 \times 10^{-1}$$
$$pH = 1.00$$

Solved Problem 13.5 Calculate the pH of 0.250 M NH_3. ($K_b = 1.8 \times 10^{-5}$.)

Solution:

	NH_3	+	H_2O	$\rightleftarrows$	NH_4^+	+	OH^-
Initial	0.250				0		0
Change	$-x$				x		x
Equilibrium	$0.250 - x$				x		x

$$K_b = \frac{[NH_4^+][OH^-]}{[NH_3]} = \frac{(x)(x)}{0.250} = 1.8 \times 10^{-5}$$

$$x^2 = 4.5 \times 10^{-6}$$

$$x = 2.1 \times 10^{-3} = [OH^-]$$

$$[H_3O^+] = \frac{K_w}{[OH^-]} = \frac{1.0 \times 10^{-14}}{2.1 \times 10^{-3}} = 4.7 \times 10^{-12}$$

$$pH = 11.33$$

Solved Problem 13.6 What chemicals are left in solution after 0.100 mol of NaOH is added to 0.200 mol of NH_4Cl?

Solution: The balanced equation is

$$NaOH + NH_4Cl \rightarrow NH_3 + NaCl + H_2O$$

The limiting quantity is NaOH, so that 0.100 mol NaOH reacts with 0.100 mol NH_4Cl to produce 0.100 mol NH_3 + 0.100 mol NaCl + 0.100 mol H_2O). There is also 0.100 mol excess NH_4Cl left in the solution. This is a buffer solution of NH_3 plus NH_4^+ in which the Na^+ and Cl^- are inert.

Chapter 14
ORGANIC CHEMISTRY

IN THIS CHAPTER:

- ✔ *Introduction*
- ✔ *Bonding in Organic Compounds*
- ✔ *Structural and Line Formulas*
- ✔ *Hydrocarbons*
- ✔ *Isomerism*
- ✔ *Radicals and Functional Groups*
- ✔ *Classes of Organic Compounds*
- ✔ *Solved Problems*

Introduction

Historically, the term **organic chemistry** has been associated with the study of compounds obtained from plants and animals. In modern terms, an **organic compound** is one that contains at least one carbon-to-carbon and/or carbon-to-hydrogen bond. (Urea and thiourea are compounds considered to be organic that do not fit this description.) In addition to carbon and hydrogen, the elements that are most likely to be present in organic compounds are oxygen, nitrogen, phosphorus, sulfur, and the halogens. With just these few elements, literally millions of organic compounds are known, and thousands of new compounds are synthesized every year.

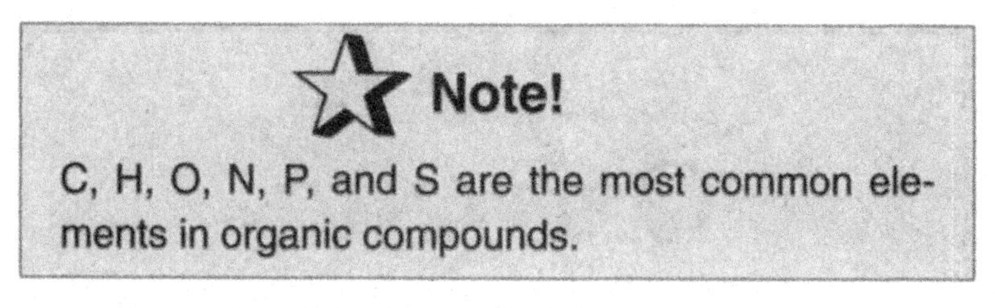

Note!

C, H, O, N, P, and S are the most common elements in organic compounds.

Bonding in Organic Compounds

The elements that are commonly part of organic compounds are all located in the upper right corner of the periodic table. They are all nonmetals. The bonds between the elements are essentially covalent. Though some organic molecules may form ions, the bonds within each organic ion are covalent.

The carbon atom has four electrons in its outermost shell. In order to complete its octet, each atom must share a total of four electron pairs. The **order** of a bond is the number of electron pairs shared in that bond. The total number of shared pairs is called the **total bond order** of an atom. Thus, carbon must have a total bond order of 4 (except in CO). A **single bond** is a sharing of one pair of electrons; a **double bond**, two; and a **triple bond**, three. Therefore, in organic compounds, each carbon atom forms four single bonds, a double bond and two single bonds, a triple bond and a single bond, or two double bonds.

A hydrogen atom has only one electron in its outermost shell, and can accommodate a maximum of two electrons in its outermost shell. Hence, in any molecule, each hydrogen atom can form only one bond, a single bond. The oxygen atom, with six electrons in its outermost shell, can complete its octet by forming either two single bonds or one double bond, for a total bond order of two (except in CO). The total bond orders of the other elements usually found in organic compounds can be deduced in a similar manner.

Structural and Line Formulas

Electron dot formulas are useful for deducing the structures of organic molecules, but it is more convenient to use simpler representations. In

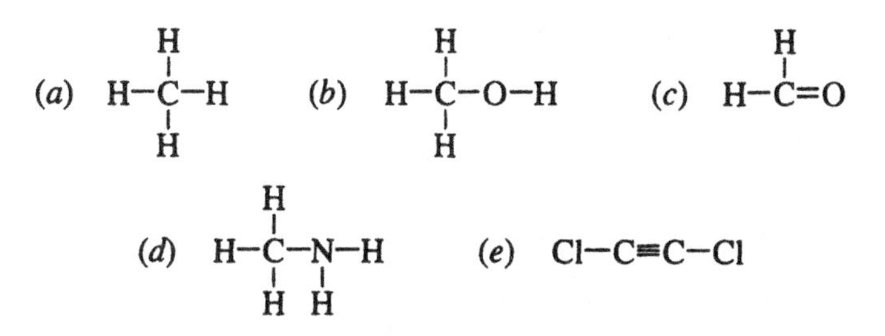

Figure 14-1 Structural formulas of (*a*) CH_4, (*b*) CH_4O, (*c*) CH_2O, (*d*) CH_5N, and (*e*) C_2Cl_2

structural or **graphical formulas**, a line is used to denote a shared pair of electrons. Because each pair of electrons shared between two atoms is equivalent to a total bond order of one, each shared pair can be represented by a line between the symbols of the elements. Unshared electrons on the atoms are often not shown in this kind of representation. The structural formulas for several compounds are shown in Figure 14-1.

For even greater convenience, **line formulas** may be used. In line formulas, the symbol for each carbon atom is written on a line adjacent to the symbols for the other elements to which it is bonded. The line formulas for the compounds shown in Figure 14-1 are (*a*) CH_4, (*b*) CH_3OH, (*c*) CH_2O, (*d*) CH_3NH_2, and (*e*) CClCCl or ClCCCl.

Hydrocarbons

Hydrocarbons consist of only carbon and hydrogen atoms. There are four main series of hydrocarbons, based on their characteristic structures: alkanes, alkenes, alkynes, and aromatic series.

The **alkane series** is also called the **saturated hydrocarbon series** because the molecules of this class have carbon atoms connected by single carbon bonds only and therefore have the maximum number of hydrogen atoms possible for the number of carbon atoms. The line formulas and names of the first ten members in the series, given in Table 14.1, should be memorized because these names form the basis for naming many other organic compounds. Note that the characteristic ending of each name is **–ane.** At room temperature, the first four members of this series are gases; the remainder of those listed in Table 14.1 are liquids;

Number of C Atoms	Molecular Formula	Line Formula	Name
1	CH_4	CH_4	Methane
2	C_2H_6	CH_3CH_3	Ethane
3	C_3H_8	$CH_3CH_2CH_3$	Propane
4	C_4H_{10}	$CH_3CH_2\ CH_2CH_3$	Butane
5	C_5H_{12}	$CH_3CH_2\ CH_2\ CH_2CH_3$	Pentane
6	C_6H_{14}	$CH_3CH_2\ CH_2\ CH_2\ CH_2CH_3$	Hexane
7	C_7H_{16}	$CH_3\ CH_2\ CH_2\ CH_2\ CH_2CH_2CH_3$	Heptane
8	C_8H_{18}	$CH_3CH_2\ CH_2\ CH_2\ CH_2\ CH_2\ CH_2CH_3$	Octane
9	C_9H_{20}	$CH\ CH_2CH_2\ CH_2\ CH_2\ CH_2\ CH_2\ CH_2CH_3$	Nonane
10	$C_{10}H_{22}$	$CH_3CH_2\ CH_2\ CH_2\ CH_2\ CH_2\ CH_2\ CH_2\ CH_2CH_3$	Decane

Table 14.1 The simplest alkanes

members having more than 13 carbon atoms are solids at room temperature.

Alkanes are rather inert chemically. Aside from burning in air or oxygen to produce carbon dioxide and water (or carbon monoxide and water), the most characteristic reaction they undergo is reaction with halogen molecules, initiated with light. For example,

$$C_5H_{12} + Br_2 \xrightarrow{\text{light}} C_5H_{11}Br + HBr$$

Because of their limited reactivity, the saturated hydrocarbons are also called the **paraffins.**

The **alkene series** of hydrocarbons is characterized by having one or more double bonds in the carbon chain of each molecule. Since there must be at least two carbon atoms present to have a carbon-to-carbon double bond, the first member of this series is ethane, C_2H_4, also known as ethylene. Propene (propylene), C_3H_6, and butane (butylenes), C_4H_8, are the next two members of the series. The systematic names of these compounds denote the number of carbon atoms in the chain with the name derived from that of the alkane having the same number of carbon atoms. Note the characteristic ending **–ene.**

Owing to the presence of the double bond, the alkenes are said to be unsaturated and are more reactive than the alkanes. The alkenes may react with hydrogen gas in the presence of a catalyst to produce the corresponding alkane. They may react with halogens or hydrogen halides at low temperatures to form compounds containing only single bonds.

These possibilities are illustrated in the following equations in which ethylene is used as a typical alkene:

$$CH_2{=}CH_2 + H_2 \xrightarrow{\text{catalyst}} CH_3CH_3$$
$$CH_2{=}CH_2 + Br_2 \rightarrow CH_2BrCH_2Br$$
$$CH_2{=}CH_2 + HBr \rightarrow CH_3CH_2Br$$

The **alkyne series** is characterized by molecules with one or more triple bonds each. They are quite reactive. Ethyne is commonly known as **acetylene**, a commercially important fuel and raw material in the manufacture of rubber and other industrial chemicals.

The **aromatic hydrocarbons** contain six-member rings of carbon atoms with each carbon attached to a maximum of one hydrogen atom. The simplest member is benzene, C_6H_6. The structural formula can be written as follows:

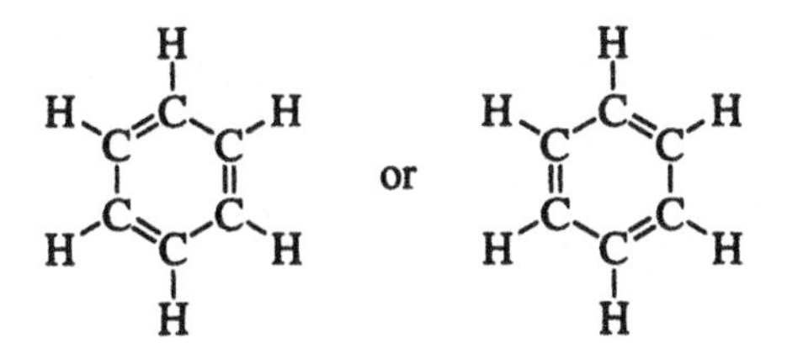

Such a molecule, containing alternating single and double bonds, would be expected to be quite reactive. Actually, it is quite unreactive due to **delocalized double bonds**. That is, the second pair of electrons in each of the three possible carbon-to-carbon double bonds is shared by all six carbon atoms rather than by any two specific carbon atoms. Two ways of writing structural formulas that indicate this type of bonding in the benzene molecules are as follows:

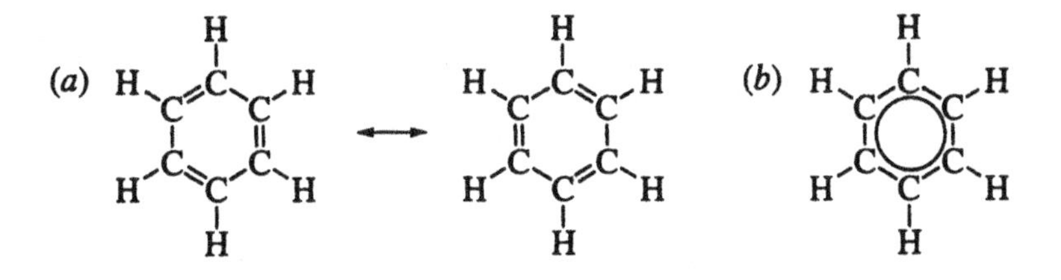

For simplicity, the ring is sometimes represented as a hexagon, each corner of which is assumed to be occupied by a carbon atom with a hydrogen atom attached (unless some other atom is explicitly indicated at that point). The delocalized electrons are indicated by a circle within the hexagon. Double or multi-ringed structures may share two carbons, and sometimes hydrocarbon chains are attached to a ringed structure.

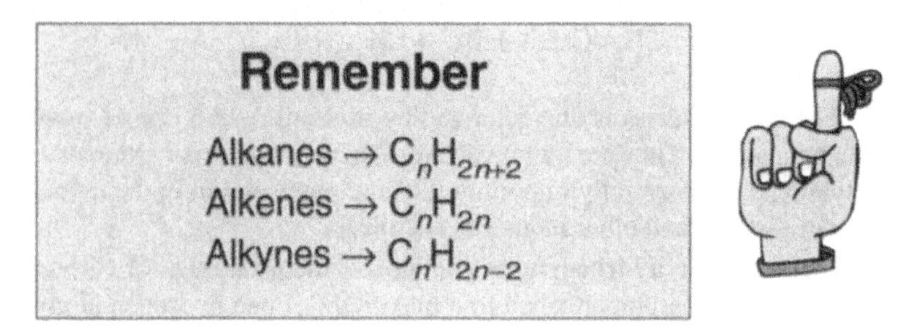

Isomerism

The ability of a carbon atom to link to more than two other carbon atoms makes it possible for two or more compounds to have the same molecular formula but different structures. Sets of compounds related in this way are called **isomers** of each other. For example, there are two different compounds having the molecular formula C_4H_{10}. Their structural formulas are as follows:

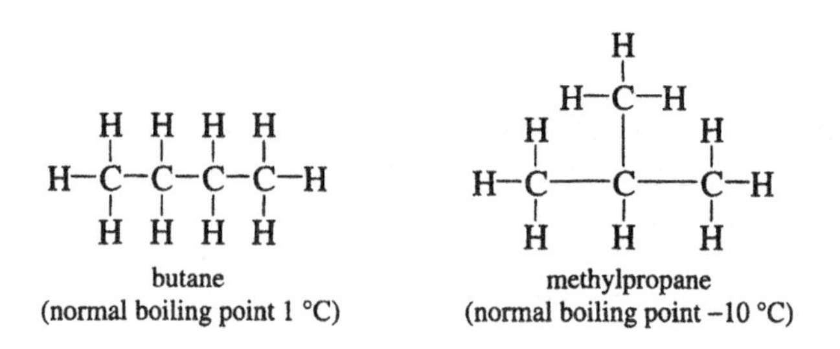

butane (normal boiling point 1 °C)

methylpropane (normal boiling point −10 °C)

These are two distinctly different compounds, with different chemical and physical properties; for example, they have different boiling points. Similarly, three isomers of pentane, C_5H_{12} exist, and so on.

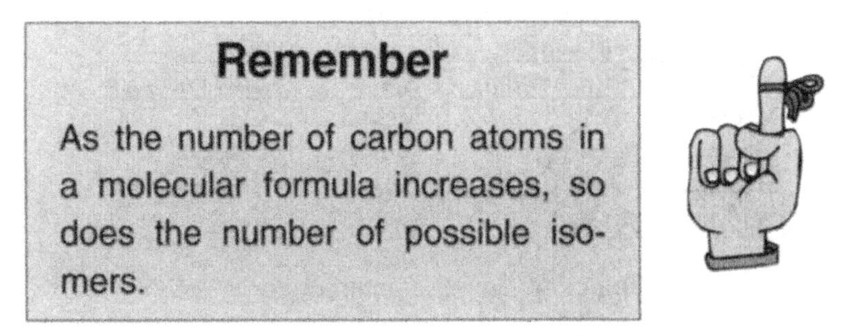

Compounds called **cycloalkanes**, having molecules with no double bonds but having a cyclic or ring structure, are isomeric with alkenes whose molecules contain the same number of carbon atoms. For example, cyclopentane and 2-pentene have the same molecular formula, C_5H_{10}, but have completely different structures:

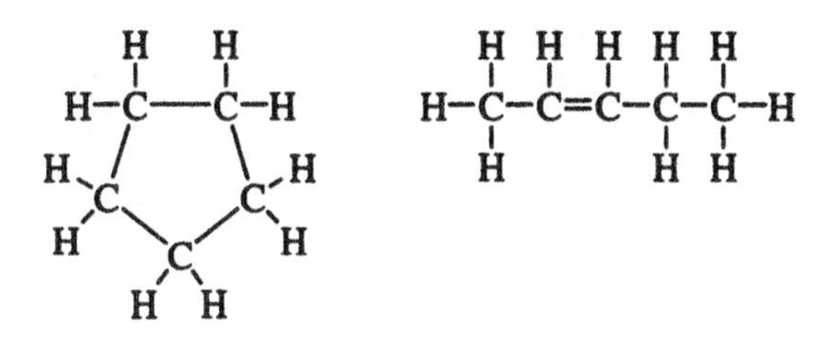

Ring structures containing double bonds, called **cycloalkenes**, can be shown to be isomeric with alkynes.

Radicals and Functional Groups

The millions of organic compounds other than hydrocarbons can be regarded as **derivatives** of hydrocarbons, where one (or more) of the hydrogen atoms on the parent molecule is replaced by another kind of atom or group of atoms. The compound is often named in a manner that designates the hydrocarbon parent from which it was derived. The hydrocarbon part of the molecule is often called the **radical** and is denoted **R** in formulas. The names of some common radicals are listed in Table 14.2.

Parent Hydrocarbon		Radical	
Name	Line Formula	Name	Line Formula
Methane	CH_4	Methyl	CH_3-
Ethane	CH_3CH_3	Ethyl	CH_3CH_2-
Propane	$CH_3CH_2CH_3$	Propyl	$CH_3CH_2CH_2-$
Benzene	C_6H_6	Phenyl	C_6H_5-

Table 14.2 Some common radicals

Since in many reactions, the hydrocarbon part of the organic compound is not changed and does not affect the nature of the reaction, it is useful to generalize many reactions using the symbol R– to denote any radical.

Type	Characteristic Functional Group	Example
Alcohol	—OH	R—OH
Ether	—O—	R—O—R′
Aldehyde	—C(—H)=O	R—C(—H)=O
Ketone	—C(=O)—	R—C(=O)—R′
Acid	—C(=O)—OH	R—C(=O)—OH
Ester	—C(=O)—O—	R—C(=O)—OR′
Amine	$—NH_2$, >NH, —N<	$R—NH_2$, RNHR′, RR′R″N

The radicals labeled R′ or R″ may be the same as or different from the radicals labeled R in the same compounds.

Table 14.3 Formulas for functional groups

When a hydrogen atom of an alkane or aromatic hydrocarbon molecule is replaced by another atom or group of atoms, the hydrocarbon-like part of the molecule is relatively inert. Therefore, the resulting compound will have properties characteristic of the substituting group. Specific groups of atoms responsible for the characteristic properties of the compound are called **functional groups**. For the most part, organic compounds can be classified according to the functional group they contain (see Table 14.3). The most important classes of such compounds include: (1) alcohols, (2) ethers, (3) aldehydes, (4) ketones, (5) acids, (6) amines, and (7) esters.

Classes of Organic Compounds

Alcohols. Compounds containing the functional group –OH are called **alcohols**. The –OH group is covalently bonded to a carbon atom in an alcohol molecule, and the molecules do not ionize in water solution to give OH^- ions. They react with metallic sodium to liberate hydrogen in a reaction analogous to that of sodium with water.

$$2\ ROH + 2\ Na \rightarrow H_2 + 2\ NaOR$$

The simplest alcohol is methanol, CH_3OH, also called methyl alcohol. Ethanol, CH_3CH_2OH, also known as ethyl alcohol, is the next simplest, containing two carbon atoms. When the number of carbon atoms in an alcohol molecule is greater than two several isomers are possible, depending on the location of the –OH group as well as on the nature of the carbon chain.

Did You Know?

Ethanol is the principal constituent of intoxicating beverages.

Ethers. A chemical reaction that removes a molecule of water from two alcohol molecules results in the formation of an **ether**.

$$CH_3CH_2OH + HOCH_2CH_3 \rightarrow CH_3CH_2OCH_2CH_3 + H_2O$$

This reaction is run by heating the alcohol in the presence of concentrated sulfuric acid, a good dehydrating agent. The radicals of the ether molecule need not be identical. A mixed ether is named after both radicals.

Aldehydes and Ketones. Aldehydes are produced by the mild oxidation of primary alcohols.

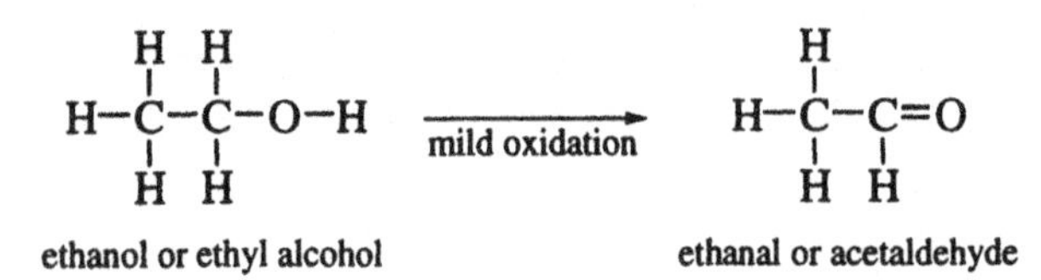

The mild oxidation of a secondary alcohol produces a **ketone**.

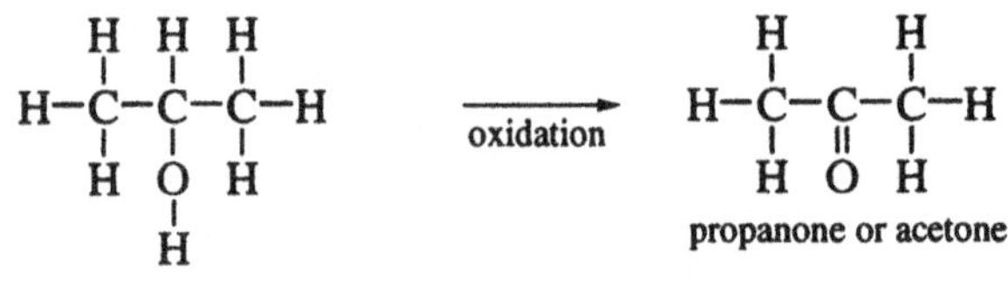

Each of these groups is characterized by having a **carbonyl group**.

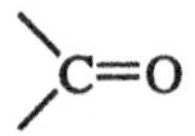

The aldehyde has the carbonyl group on one end of the carbon chain, and the ketone has the carbonyl group on a carbon other than one on the end. The systematic name ending for aldehydes is –**al**; that for ketones is –**one**.

Acids and Esters. Acids can be produced by the oxidation of aldehydes or primary alcohols. For example, ethyl alcohol oxidizes to acetic acid (also known as ethanoic acid or vinegar):

$$CH_3CH_2OH \xrightarrow[\text{oxidation}]{} CH_3COOH$$

Acids react with alcohols to produce **esters**. For example, acetic acid reacts with ethyl alcohol to produce the ester ethyl acetate and water:

$$CH_3COOH + HOCH_2CH_3 \rightarrow CH_3COOCH_2CH_3 + H_2O$$

Esters are named by combining the radical name of the alcohol with that of the negative ion of the acid. The ending **–ate** replaces the **–oic** acid of the parent acid. The formation of an ester from an alcohol and an acid is an equilibrium reaction. The reverse reaction can be promoted by removing the acid from the reaction mixture.

Amines. **Amines** can be considered derivatives of ammonia, NH_3, in which one or more hydrogen atoms have been replaced by organic radicals. For example, replacing one hydrogen atom on the nitrogen atom results in a **primary amine**, RNH_2. A **secondary amine** has a formula of the type R_2NH, and a **tertiary amine** has no hydrogen atoms on the nitrogen atom in its molecules and has the formula R_3N. Like ammonia, amines react as Brønsted bases:

$$RNH_2 + H_2O \rightleftarrows RNH_3^+ + OH^-$$

Solved Problems

Solved Problem 14.1 What is the total bond order of carbon in (*a*) CO_2, (*b*) CO, (*c*) $H_2C{=}O$, and (*d*) CH_4?

Solution: (*a*) 4, (*b*) 3, (*c*) 4, and (*d*) 4.

Solved Problem 14.2 Write the structural formulas for the molecules represented by the following formulas: (*a*) C_2H_4 and (*b*) CH_3COCH_3.

Solution: (*a*) Since the hydrogen atoms can have only a total bond order of 1, the two carbon atoms must be linked together. In order for each carbon atom to have a total bond order of 4, the two carbon atoms must be linked to each other by a double bond and also be bonded to two hydrogen atoms each.

```
   H  H
   |  |
 H—C=C—H
```

(*b*) The line formula CH_3COCH_3 implies that two of the three carbon atoms each have three hydrogen atoms attached. This permits them to form one additional single bond, to the middle carbon atom. The middle carbon atom, with two single bonds to carbon atoms, must complete its total bond order of 4 with a double bond to the oxygen atom.

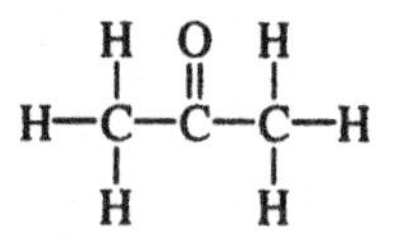

Solved Problem 14.3 Write the structural formulas for (*a*) benzene, (*b*) toluene (methyl benzene), and (*c*) bromobenzene.

Solution:

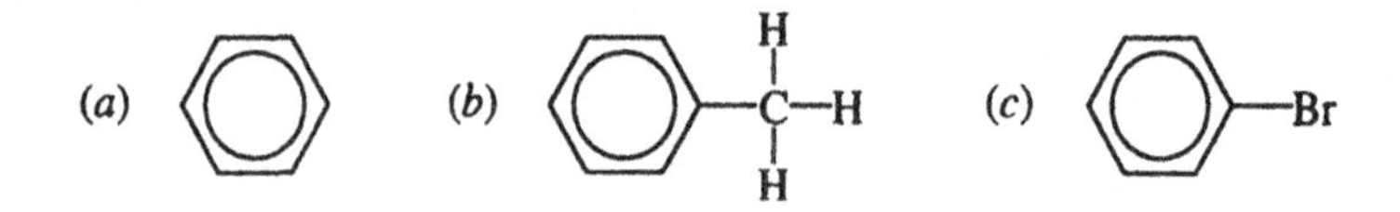

Solved Problem 14.4 Write the structural formulas for the three isomers of pentane.

Solution:

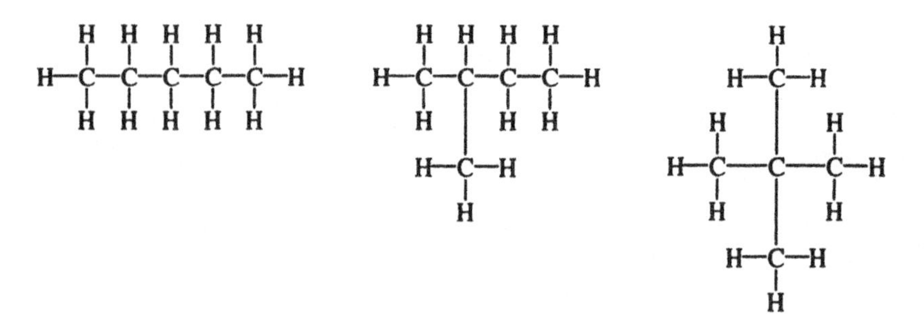

Solved Problem 14.5 Name the following ethers: (*a*) $CH_3OCH_2CH_2CH_3$, (*b*) $C_6H_5OCH_3$, and (*c*) CH_3OCH_3.

Solution: (*a*) Methyl propyl ether, (*b*) phenyl methyl ether, and (*c*) dimethyl ether.

Solved Problem 14.6 Name the following esters: (*a*) $C_4H_9OCOCH_3$ and (*b*) C_6H_5OCOH.

Solution: (*a*) Butyl acetate and (*b*) phenyl formate.

Solved Problem 14.7 Write the formulas for (*a*) ammonium chloride, (*b*) methyl ammonium chloride, (*c*) dimethyl ammonium chloride, (*d*) trimethyl ammonium chloride, and (*e*) tetramethyl ammonium chloride.

Solution: (*a*) NH_4Cl. (*b*) CH_3NH_3Cl. One H atom has been replaced by a CH_3 group. (*c*) $(CH_3)_2NH_2Cl$. Two H atoms have been replaced by CH_3 groups. (*d*) $(CH_3)_3NHCl$. Three H atoms have been replaced by CH_3 groups. (*e*) $(CH_3)_4NCl$. All four H atoms have been replaced by CH_3 groups.

APPENDIX

Periodic Table

Group Numbers																		
Classical	IA	IIA	IIIB	IVB	VB	VIB	VIIB	VIII			IB	IIB	IIIA	IVA	VA	VIA	VIIA	0
Amended	IA	IIA	IIIA	IVA	VA	VIA	VIIA	VIII			IB	IIB	IIIB	IVB	VB	VIB	VIIB	0
Modern	1	2	3	4	5	6	7	8	9	10	11	12	13	14	15	16	17	18
	1 H 1.0080																	2 He 4.00260
	3 Li 6.941	4 Be 9.01218											5 B 10.81	6 C 12.011	7 N 14.0067	8 O 15.9994	9 F 18.9984	10 Ne 20.179
	11 Na 22.9898	12 Mg 24.305											13 Al 26.9815	14 Si 28.086	15 P 30.9738	16 S 32.06	17 Cl 35.453	18 Ar 39.948
	19 K 39.102	20 Ca 40.08	21 Sc 44.9559	22 Ti 47.90	23 V 50.9414	24 Cr 51.996	25 Mn 54.9380	26 Fe 55.847	27 Co 58.9332	28 Ni 58.71	29 Cu 63.546	30 Zn 65.37	31 Ga 69.72	32 Ge 72.59	33 As 74.9216	34 Se 78.96	35 Br 79.904	36 Kr 83.80
	37 Rb 85.4678	38 Sr 87.62	39 Y 88.9059	40 Zr 91.22	41 Nb 92.9064	42 Mo 95.94	43 Tc 98.9062	44 Ru 101.07	45 Rh 102.9055	46 Pd 106.4	47 Ag 107.868	48 Cd 112.40	49 In 114.82	50 Sn 118.69	51 Sb 121.75	52 Te 127.60	53 I 126.9045	54 Xe 131.30
	55 Cs 132.9055	56 Ba 137.34	57 La 138.9055 •	72 Hf 178.49	73 Ta 180.9479	74 W 183.85	75 Re 186.2	76 Os 190.2	77 Ir 192.22	78 Pt 195.09	79 Au 196.9665	80 Hg 200.59	81 Tl 204.37	82 Pb 207.2	83 Bi 208.9806	84 Po (210)	85 At (210)	86 Rn (222)
	87 Fr (223)	88 Ra 226.0254	89 Ac (227) †	104 Unq (261)	105 Unp (262)	106 Unh (263)												

•	58 Ce 140.12	59 Pr 140.9077	60 Nd 144.24	61 Pm (145)	62 Sm 150.4	63 Eu 151.96	64 Gd 157.25	65 Tb 158.9254	66 Dy 162.50	67 Ho 164.9303	68 Er 167.26	69 Tm 168.9342	70 Yb 173.04	71 Lu 174.97
†	90 Th 232.0381	91 Pa 231.0359	92 U 238.029	93 Np 237.0482	94 Pu (242)	95 Am (243)	96 Cm (247)	97 Bk (249)	98 Cf (251)	99 Es (254)	100 Fm (253)	101 Md (256)	102 No (254)	103 Lr (257)

Index

CPSIA information can be obtained at www.ICGtesting.com
Printed in the USA
BVOW06s1128110816

458587BV00014B/81/P